Equity-Linked Life Insurance
Partial Hedging Methods

CHAPMAN & HALL/CRC
Financial Mathematics Series

Aims and scope:
The field of financial mathematics forms an ever-expanding slice of the financial sector. This series aims to capture new developments and summarize what is known over the whole spectrum of this field. It will include a broad range of textbooks, reference works and handbooks that are meant to appeal to both academics and practitioners. The inclusion of numerical code and concrete real-world examples is highly encouraged.

Series Editors

M.A.H. Dempster
Centre for Financial Research
Department of Pure
Mathematics and Statistics
University of Cambridge

Dilip B. Madan
Robert H. Smith School
of Business
University of Maryland

Rama Cont
Department of Mathematics
Imperial College

Published Titles

American-Style Derivatives; Valuation and Computation, *Jerome Detemple*

Analysis, Geometry, and Modeling in Finance: Advanced Methods in Option Pricing, *Pierre Henry-Labordère*

C++ for Financial Mathematics, *John Armstrong*

Commodities, *M. A. H. Dempster and Ke Tang*

Computational Methods in Finance, *Ali Hirsa*

Counterparty Risk and Funding: A Tale of Two Puzzles, *Stéphane Crépey and Tomasz R. Bielecki, With an Introductory Dialogue by Damiano Brigo*

Credit Risk: Models, Derivatives, and Management, *Niklas Wagner*

Engineering BGM, *Alan Brace*

Equity-Linked Life Insurance: Partial Hedging Methods, *Alexander Melnikov and Amir Nosrati*

Financial Mathematics: A Comprehensive Treatment, *Giuseppe Campolieti and Roman N. Makarov*

The Financial Mathematics of Market Liquidity: From Optimal Execution to Market Making, *Olivier Guéant*

Financial Modelling with Jump Processes, *Rama Cont and Peter Tankov*

Interest Rate Modeling: Theory and Practice, *Lixin Wu*

Introduction to Credit Risk Modeling, Second Edition, *Christian Bluhm, Ludger Overbeck, and Christoph Wagner*

An Introduction to Exotic Option Pricing, *Peter Buchen*

Introduction to Risk Parity and Budgeting, *Thierry Roncalli*

Introduction to Stochastic Calculus Applied to Finance, Second Edition, *Damien Lamberton and Bernard Lapeyre*

Model-Free Hedging: A Martingale Optimal Transport Viewpoint,
  Pierre Henry-Labordère
Monte Carlo Methods and Models in Finance and Insurance, *Ralf Korn, Elke Korn,
  and Gerald Kroisandt*
Monte Carlo Simulation with Applications to Finance, *Hui Wang*
Nonlinear Option Pricing, *Julien Guyon and Pierre Henry-Labordère*
Numerical Methods for Finance, *John A. D. Appleby, David C. Edelman,
  and John J. H. Miller*
Option Valuation: A First Course in Financial Mathematics, *Hugo D. Junghenn*
Portfolio Optimization and Performance Analysis, *Jean-Luc Prigent*
Quantitative Finance: An Object-Oriented Approach in C++, *Erik Schlögl*
Quantitative Fund Management, *M. A. H. Dempster, Georg Pflug,
  and Gautam Mitra*
Risk Analysis in Finance and Insurance, Second Edition, *Alexander Melnikov*
Robust Libor Modelling and Pricing of Derivative Products, *John Schoenmakers*
Stochastic Finance: An Introduction with Market Examples, *Nicolas Privault*
Stochastic Finance: A Numeraire Approach, *Jan Vecer*
Stochastic Financial Models, *Douglas Kennedy*
Stochastic Processes with Applications to Finance, Second Edition,
  Masaaki Kijima
Stochastic Volatility Modeling, *Lorenzo Bergomi*
Structured Credit Portfolio Analysis, Baskets & CDOs, *Christian Bluhm
  and Ludger Overbeck*
Understanding Risk: The Theory and Practice of Financial Risk Management,
  David Murphy
Unravelling the Credit Crunch, *David Murphy*

Proposals for the series should be submitted to one of the series editors above or directly to:
CRC Press, Taylor & Francis Group
3 Park Square, Milton Park
Abingdon, Oxfordshire OX14 4RN
UK

Chapman & Hall/CRC FINANCIAL MATHEMATICS SERIES

Equity-Linked Life Insurance
Partial Hedging Methods

Alexander Melnikov
Amir Nosrati

CRC Press
Taylor & Francis Group
Boca Raton London New York

CRC Press is an imprint of the
Taylor & Francis Group, an **informa** business

A CHAPMAN & HALL BOOK

CRC Press
Taylor & Francis Group
6000 Broken Sound Parkway NW, Suite 300
Boca Raton, FL 33487-2742

First issued in paperback 2020

© 2018 by Taylor & Francis Group, LLC
CRC Press is an imprint of Taylor & Francis Group, an Informa business

No claim to original U.S. Government works

ISBN-13: 978-1-4822-4026-9 (hbk)
ISBN-13: 978-0-367-65777-2 (pbk)

This book contains information obtained from authentic and highly regarded sources. Reasonable efforts have been made to publish reliable data and information, but the author and publisher cannot assume responsibility for the validity of all materials or the consequences of their use. The authors and publishers have attempted to trace the copyright holders of all material reproduced in this publication and apologize to copyright holders if permission to publish in this form has not been obtained. If any copyright material has not been acknowledged please write and let us know so we may rectify in any future reprint.

Except as permitted under U.S. Copyright Law, no part of this book may be reprinted, reproduced, transmitted, or utilized in any form by any electronic, mechanical, or other means, now known or hereafter invented, including photocopying, microfilming, and recording, or in any information storage or retrieval system, without written permission from the publishers.

For permission to photocopy or use material electronically from this work, please access www.copyright.com (http://www.copyright.com/) or contact the Copyright Clearance Center, Inc. (CCC), 222 Rosewood Drive, Danvers, MA 01923, 978-750-8400. CCC is a not-for-profit organization that provides licenses and registration for a variety of users. For organizations that have been granted a photocopy license by the CCC, a separate system of payment has been arranged.

Trademark Notice: Product or corporate names may be trademarks or registered trademarks, and are used only for identification and explanation without intent to infringe.

Visit the Taylor & Francis Web site at
http://www.taylorandfrancis.com

and the CRC Press Web site at
http://www.crcpress.com

Contents

Preface		**ix**
1	**Basic notions and facts from stochastic analysis, mathematical finance and insurance**	**1**
	1.1 Stochastic analysis	1
	1.2 Contemporary stochastic analysis	6
	1.3 Mathematical finance and insurance	13
2	**Quantile hedging of equity-linked life insurance contracts in the Black–Scholes model**	**23**
	2.1 Quantile hedging of contracts with deterministic guarantees	23
	2.2 Quantile hedging of contracts with stochastic guarantees	32
	2.3 Quantile portfolio management with multiple contracts with a numerical example	46
3	**Valuation of equity-linked life insurance contracts via efficient hedging in the Black–Scholes model**	**53**
	3.1 Efficient hedging and pricing formulas for contracts with correlated assets and guarantees	53
	3.2 Efficient hedging and pricing of contracts with perfectly correlated stocks and guarantees	64
	3.3 Illustrative risk-management for the risk-taking insurer	72
4	**Quantile hedging and risk management of contracts for diffusion and jump-diffusion models**	**77**
	4.1 Quantile pricing formulas for contracts with perfectly correlated stocks and guarantees: Diffusion and jump-diffusion settings	77
	4.2 Quantile hedging and an exemplary risk-management scheme for equity-linked life insurance	90

5 CVaR-Hedging: Theory and applications — 107

- 5.1 CVaR-hedging methodology and general theoretical facts .. 107
- 5.2 CVaR-hedging in the Black–Scholes model with applications to equity-linked life insurance 115
- 5.3 CVaR-hedging in the regime-switching telegraph market model 125

6 Defaultable securities and equity-linked life insurance contracts — 137

- 6.1 Multiple defaults and defaultable claims in the Black–Scholes model 137
- 6.2 Efficient hedging in defaultable market: Essence of the technique and main results 143
- 6.3 Application to equity-linked life insurance contracts: Brennan–Schwartz approach vs. the superhedging 158

7 Equity-linked life insurance contracts and Bermudan options — 163

- 7.1 GMDB life insurance contracts as Bermudan options 163
- 7.2 Quantile hedging of GMDB life insurance contracts 170
- 7.3 Numerical illustrations 181

Bibliographic Remarks — 187

Bibliography — 193

Index — 199

Preface

The dynamics of financial systems have changed tremendously since the beginning of the 1970s due to financial innovations and new quantitative methods. The insurance industry as an integral part of the financial system has grown at a tremendous pace too. Correspondingly, equity-linked insurance has been especially successful and become a popular subject of research and industrial applications. Innovative insurance instruments, also known as index-linked insurance, variable annuities and segregated funds, combine both financial and insurance risks and allow insurance structures to stay competitive in the modern financial system. It was necessary to incorporate a new randomness into actuarial calculations, stemming from the financial randomness of insurance guarantees. Thus, actuarial theory and practice had to be enriched by contemporary techniques and methods of mathematical finance. In life insurance there appeared contracts in which the payoff was connected to the market value of risky assets and conditioned by the survival status of the insured. It is equity-linked life insurance or insurance linked to risky assets of the financial markets. Therefore, it is not a surprise that the famous Black–Scholes formula appears in many actuarial calculations as their integral part.

Certainly, there is a need for books primarily devoted to this motivating and now well-developed topic. In this regard, the monograph "Investment guarantees: Modeling and risk management for equity-linked life insurance" by M. Hardy can be mentioned as the first book where such contracts are systematically investigated with the help of stochastic modeling.

Our book presents much more advanced mathematical methods and techniques developed in the last two decades in mathematical finance and actuarial science. The book presents existing hedging methodologies in order to optimally price and hedge equity-linked life insurance contracts whose pay-off depends on the performance of risky assets. Such contracts incorporate financial risk, which stems from the uncertainty about future prices of the underlying assets, and insurance risk, which arises from the policy-holder's mortality. The book contains a number of theoretical results, formulas, numerical examples and illustrative materials for pricing and hedging of equity-linked life insurance contracts. The essence of our approach is partial/imperfect hedging (quantile hedging, efficient hedging, conditional value-at-risk (CVaR) hedging), developed in this context for different financial market models (Black–Scholes model, two-factor diffusion and jump-diffusion models, regime-switching model) and for deterministic and stochastic insurance guarantees.

The book contains seven chapters written in a methodical homogeneous manner, which demonstrate the relationship between theoretical findings and their practical implementations. Chapter 1 creates a necessary and useful background from mathematical finance, life insurance, and stochastic analysis. Chapters 2 and 3 are devoted to quantile and efficient hedging of equity-linked life insurance contracts in the framework of the Black–Scholes model. Chapter 4 contains a variety of pricing formulas for quantile hedging for diffusion and jump-diffusion market models and for insurance contracts with constant and flexible guarantees. Based on these results a convenient scheme of quantile risk management is given as well as its numerical realization. Chapter 5 deals with the CVaR hedging of contracts in the Black–Scholes model and in a regime-switching market model. Chapter 6 states a natural connection between defaultable contingent claims and equity-linked life insurance contracts. Combining such a connection with the super-hedging technique, the problem of efficient hedging of pure endowment life insurance contracts is solved and numerically illustrated. The contracts considered in Chapters 2–6 belong to a European type of options. Chapter 7 shows how a popular class of equity-linked life insurance contracts, GMMB (guaranteed minimum maturity benefit) and GMDB (guaranteed minimum death benefit), are naturally connected with Bermudan options, a specific type of American option. Based on this nice observation, the explicit pricing formulas of these contracts are derived with the help of a fruitful combination of quantile hedging and super-hedging techniques.

Equity-linked life insurance is now a rather wide area of research and application. Therefore, it is difficult to include all the exploited methods without loss of the homogeneous character of a particular book. This is a reason why the well-developed approaches (quadratic hedging and utility maximization) are not presented in the monograph.

We hope that our book will be very useful for different categories of readers, from graduate students to professors, experts and practitioners working in the area of life insurance and finance.

Alexander Melnikov
Amir Nosrati
University of Alberta
Edmonton, AB, Canada

Chapter 1

Basic notions and facts from stochastic analysis, mathematical finance and insurance

1.1	Stochastic analysis	1
1.2	Contemporary stochastic analysis	6
1.3	Mathematical finance and insurance	13

1.1 Stochastic analysis

A fundamental concept in contemporary stochastic analysis is the stochastic basis. This is a complete probability space $(\Omega, \mathcal{F}, \mathrm{P})$ endowed with a filtration $\mathbb{F} = (\mathcal{F}_t)_{t \geq 0}$, which is a nondecreasing right-continuous family of σ-algebras. $\mathcal{F}_t \subseteq \mathcal{F}$ completed by P-null sets from $\mathcal{F}$. As usual, we assume that all stochastic processes $X = (X_t(\omega))_{t \geq 0}$ under consideration are defined on the stochastic basis $(\Omega, \mathcal{F}, \mathbb{F}, \mathrm{P})$ and are $\mathcal{F}$-adapted (for each fixed t, the random variable X_t is measurable with respect to $\mathcal{F}_t$). Moreover, for almost all $\omega \in \Omega$, the trajectories $X.(\omega)$ belong to the space $\mathbb{D}(\mathbb{R})$ of functions $x(t)$, $t \geq 0$, right continuous with finite left limits, taking the values in a space $\mathbb{R}$. As a rule, this space will be the d-dimensional Euclidean space $\mathbb{R}^d$, $d \geq 1$. The measurability of X is understood as the measurability of the mapping:

$$X : \Omega \times \mathbb{R}_+ \to \mathbb{R},$$

with respect to the σ-algebras $\mathcal{F} \times \mathcal{B}(\mathbb{R}_+)$ and $\mathcal{B}(\mathbb{R})$, where $\mathbb{R}_+ = [0, \infty)$, and $\mathcal{B}(\cdot)$ are the Borel σ-algebras. The progressive measurability of X is understood as the measurability of the mapping:

$$X : (\Omega \times [0, t], \mathcal{F}_t \times \mathcal{B}([0, t])) \to (\mathbb{R}, \mathcal{B}(\mathbb{R}))$$

for any $t \geq 0$. Two stochastic processes are said to be modifications of each other if for any $t \in \mathbb{R}_+$:

$$\mathrm{P}(\omega : X_t(\omega) = Y_t(\omega)) = 1.$$

The stronger notion of the indistinguishability of X and Y is expressed as

$$\mathrm{P}(\omega : X_t(\omega) = Y_t(\omega) \text{ for all } t \in \mathbb{R}_+) = 1.$$

The continuous process W is a Wiener process, or Brownian motion, if the following conditions are satisfied:

- $W_0 = 0$ (P – a.s.);
- $W_t - W_s$ does not depend on $\mathcal{F}_s$, $s \leq t$;
- $W_t - W_s$ has a normal distribution with zero mean and variance $t - s$, that is $N(0, t - s)$.

We need to understand what the distribution of a process X is. To describe it, we denote by $\mathbb{R}^{[0,\infty)}$ the set of all possible functions $x : \mathbb{R}_+ \to \mathbb{R}$. A subset $A \subseteq \mathbb{R}^{[0,\infty)}$ is called a cylindrical set if there is an n-tuple $t_1 < ... < t_n$ and a set $B \in \mathcal{B}(\mathbb{R}^n)$ such that $A = \{x : (x_{t_1}, ..., x_{t_n}) \in B_n\}$. Such sets generate a corresponding smallest σ–algebra $\mathcal{B}^{[0,\infty)}$ containing them. The distribution of a random process X is defined to be the probability measure $\mu = \mu^X$ on $\mathcal{B}^{[0,\infty)}$ such that

$$\mu(C) = \mu^X(C) = P(\omega : X.(\omega) \in C), \ C \in \mathcal{B}^{[0,\infty)}.$$

A finite-dimensional distribution of X is the measure $\mu_{t_1...t_n} = \mu^X_{t_1...t_n}$ given by an equality

$$\mu^X_{t_1...t_n}(A) = P(\omega : (X_{t_1}(\omega), ..., X_{t_n}(\omega)) \in B_n).$$

If a cylindrical set A has another representation $A = \{x : (x_{s_1}, ..., x_{s_m}) \in B_m\}$, the equality (for any A)

$$\mu^X_{s_1...s_m}(A) = \mu^X_{t_1...t_n}(A)$$

is called the consistency condition. The famous Kolmogorov consistency theorem claims that for a given system $\{\mu_{t_1...t_n}\}$ of consistent finite-dimensional distributions, there exits a unique distribution μ with the measures $\mu_{t_1...t_n}$ as its finite-dimensional distributions on $(\mathbb{R}^{[0,\infty)}, \mathcal{B}^{[0,\infty)})$, and for the coordinate process $X_t(\omega) = \omega(t)$, $\omega \in \Omega = \mathbb{R}^{[0,\infty)}$. Its distribution μ^X coincides with the measure μ. Applying this theorem to the family of measures $\{\mu_{t_1...t_n}\}$ admitting the Gaussian densities:

$$P_{t_1...t_n}(x_1, ..., x_n) = (2\pi t_1)^{-\frac{1}{2}} e^{-\frac{x_1^2}{2t_1}} ... (2\pi(t_n - t_{n-1}))^{-\frac{1}{2}} e^{-\frac{(x_n - x_{n-1})^2}{2(t_n - t_{n-1})}}$$

we arrive at the unique distribution $\mu = \mu^W$ called the Wiener measure. The corresponding coordinate process $W_t(\omega) = \omega(t)$ is called a Wiener process. The above construction of a Wiener process is given in the space $\mathbb{R}^{[0,\infty)}$. However, its sample paths are continuous functions because of another theorem from Kolmogorov:

Assume that for a random process $(X_t)_{t \geq 0}$, there exist constants C, ε, $\beta > 0$ such that $E|X_t - X_s|^\beta \leq C|t - s|^{1+\varepsilon}$ for all $t \geq s \geq 0$. Then X admits a continuous modification.

In the case $X_t = W_t$, we have $E|W_t - W_s|^4 = 3|t-s|^2$, and therefore the conditions of this theorem hold.

Another concept of stochastic analysis and the theory of probability is connected to absolute continuity and the singularity of probability measures. The measure $\widetilde{P}$ is absolutely continuous with respect to measure P if $\widetilde{P}(A) = 0$ for any P−null set $A \in \mathcal{F}$ ($\widetilde{P} \ll P$). This property admits a very useful and important description by means of the Radon–Nikodym density Z: for any $A \in \mathcal{F}$, $\widetilde{P}(A) = \widetilde{E}\mathbb{1}_A = E\mathbb{1}_A Z$, where E and $\widetilde{E}$ are expectations with respect to P and $\widetilde{P}$ measures, $EZ = 1$, $Z \geq 0$ (a.s.). The random variable Z is called a density of $\widetilde{P}$ with respect to P. If $\widetilde{P} \ll P$ and $P \ll \widetilde{P}$, then $Z > 0$ (a.s.), and the measures P and $\widetilde{P}$ are equivalent. Applying the above procedure to the σ-algebras $\mathcal{F}_t$ and restrictions $\widetilde{P}_t$ and P_t of $\widetilde{P}$ and P to $\mathcal{F}_t$, we arrive at the process $(Z_t)_{t \geq 0}$ called a local density of $\widetilde{P}$ and P.

Another building block in our description of the theory of stochastic processes is the diffusion process, which we consider with the help of stochastic integration with respect to a Wiener process $W = (W_t, \mathcal{F}_t)_{t \geq 0}$. Let us consider a simple function

$$f(t) = \sum_{k=1}^{n} f_{k-1} \mathbb{1}_{(t_{k-1}, t_k]}(t),$$

where $0 = t_0 < t_1 < ... < t_n = T$, f_{k-1} is $\mathcal{F}_{t_{k-1}}$-measurable, $t \leq T$.

We define the stochastic integral by the sum

$$\int_0^t f(s) dW_s = \sum_{k=1}^{n} f_{k-1} \left(W_{t_k \wedge t} - W_{t_{k-1} \wedge t} \right).$$

The integral has the following properties:

- $\int_0^t (\alpha f_s + \beta g_s) dW_s = \alpha \int_0^t f_s dW_s + \beta \int_0^t g_s dW_s$, where α and β are constants, and f and g are simple functions;

- $E \left(\int_0^T f(s) dW_s \mid \mathcal{F}_t \right) = \int_0^t f(s) dW_s$;

- $E \left(\int_0^t f_s dW_s \int_0^t g_s dW_s \right) = E \int_0^t f_s g_s dW_s$.

This stochastic integral can be extended as a linear operator to the set of progressively measurable functions such that $E \int_0^t f_s^2 dW_s < 0$ with preservation of the above properties (linearity, martingality and isometry).

Using stochastic integrals one can construct the whole class of stochastic processes (Ito's processes) of the form:

$$X_t = X_0 + \int_0^t b_s(\omega)\,ds + \int_0^t \sigma_s(\omega)\,dW_s, \qquad (1.1)$$

where X_0 is a random variable, $b = (b_t(\omega))_{t \geq 0}$ is adapted and the Lebesgue integrable random process for almost all $\omega \in \Omega$, and $\sigma = (\sigma_t(\omega))_{t \geq 0}$ is a progressively measurable random function for which the corresponding stochastic integral is well-defined.

There are two types of transformations in the class of the Ito processes (1.1): changes of variables and of measures. If the change of variables is provided by the function $F = F(t,x) \in C^{1,2}$, then a new process $Y_t = F(t, X_t)$ has the following form, called the Ito formula:

$$F(t, X_t) = F(0, X_0) + \int_0^t \left[\frac{\partial F}{\partial t}(s, X_s) + \frac{\partial F}{\partial t}(s, X_s) b(s, \omega) \right.$$

$$\left. + \frac{1}{2} \frac{\partial^2 F}{\partial^2 x}(s, X_s) \sigma^2(s, \omega) \right] ds + \int_0^t \frac{\partial F}{\partial x}(s, X_s) \sigma(s, \omega)\,dW_s. \qquad (1.2)$$

The formula (1.2) shows that the class of Ito's processes is invariant with respect to such smooth transformations.

A measure change in the class of Ito's processes leads to a miracle result called the Girsanov theorem:

Consider a progressively measurable process $\beta = (\beta_t(\omega))_{t \geq 0}$ for which the Girsanov exponential process

$$Z_t = \exp\left\{ \int_0^t \beta_s\,dW_s - \frac{1}{2}\int_0^t \beta_s^2\,ds \right\}, t \leq T,$$

is well defined. Let $E Z_T = 1$, define a new measure $\widetilde{P} \sim P$ as $\frac{d\widetilde{P}}{dP} = Z_T$ and $\widetilde{W}_t = W_t - \int_0^t \beta_s^2\,ds$. Then the process $\widetilde{W} = (\widetilde{W}_t)_{t \geq 0}$ is a Wiener process with respect to $\widetilde{P}$.

To estimate how wide the class of Ito processes (1.1) is, we can start with a Wiener process W given on a stochastic basis $(\Omega, \mathcal{F}, \mathbb{F}, (\mathcal{F}_t)_{t \geq 0}, P)$ and two functions $\tilde{b} = \tilde{b}(t,x)$ and $\tilde{\sigma} = \tilde{\sigma}(t,x)$, (t,x)-measurable and Lipschitz-continuous with respect to x. In this case one can prove that there exists a unique (up to indistinguishability) $\mathbb{F}$-adapted continuous process $X = (X_t)_{t \geq 0}$ such that the following equality is fulfilled (a.s.) for each $t \geq 0$:

$$X_t(\omega) = X_0(\omega) + \int_0^t \tilde{b}_s(X_s(\omega))\,ds + \int_0^t \tilde{\sigma}_s(X_s(\omega))\,dW_s \qquad (1.3)$$

The expression (1.3) is called a stochastic differential equation with coefficients $\tilde{b}$ and $\tilde{\sigma}$, and X is its strong solution, known also as a diffusion process with the drift parameter $\tilde{b}$ and the diffusion parameter $\tilde{\sigma}$.

Let us note that the first general construction of diffusion processes was done by Kolmogorov in terms of Markov transition probability and without a stochastic integration technique. He also investigated smooth transformations of diffusion processes, and his formulas for transformed drift and diffusion coefficients are completely consistent with the Ito formula (1.2).

Just as the class of diffusion processes is exhausted by the solutions of stochastic differential equations, the class of random processes M adapted to the filtration $\mathbb{F} = \mathbb{F}^W$ generated by a Wiener process W, and such that

$$\mathrm{E}\,|M_t| < \infty \text{ and } \mathrm{E}\,(M_t|\mathcal{F}_s) = M_s \ (\text{P- a.s.}) \tag{1.4}$$

is exhausted by the stochastic integrals $\int_0^t \varphi_s dW_s$. The process M satisfying (1.4) is called a martingale. The following martingale representation theorem plays a key role in what follows.

Any martingale $M = (M_t, \mathcal{F}_t^W)_{t \geqslant 0}$ can be represented in the form

$$M_t = M_0 + \int_0^t \varphi_s dW_s,$$

where φ is a progressively measurable random process.

We remark that the properties of stochastic integrals have the martingale property built in from the start, and the representation theorem says that there are simply no other martingales with respect to $\mathbb{F} = \mathbb{F}^W$.

Besides, the Wiener process, there is another basic process $X = (X_t, \mathcal{F}_t)_{t \geqslant 0}$ known as the Poisson process with intensity $\lambda > 0$, which is defined as follows:

- $X_0 = 0$ P-a.s.;
- $X_t - X_s$ does not depend on $\mathcal{F}_s$, $s \leqslant t$;
- $X_t - X_s$ has a Poisson distribution with parameter $\lambda(t-s)$.

It is clear that $N_t = X_t - \lambda t$ is a martingale (a Poisson martingale). It turns out that any other martingale $M = (M_t, \mathcal{F}_t^N)_{t \geqslant 0}$ can be represented in the form of an integral with respect to a Poisson martingale:

$$M_t = M_0 + \int_0^t \varphi_s dN_s,$$

where φ is a progressively measurable process, and the integral can be understood as the difference

$$\int_0^t \varphi_s dN_s = \int_0^t \varphi_s dX_s - \int_0^t \varphi_s d\lambda s$$

of two Lebesgue–Stieltjes integrals, since X_t and λt are non-decreasing processes.

We give also vector forms of the Ito process (1.1) and of the Ito formula (1.2). Namely, (1.2) can be understood as a vector equation with $b_t = \left(b_t^1, ..., b_t^d\right)^*$, $\sigma_t = \left(\sigma_t^{ij}\right)_{j=1,...,d_1}^{i=1,...,d}$ and $W_t = \left(W_t^1, ..., W_t^d\right)^*$. In this case, and for a smooth function: $F: \mathbb{R}_+ \times \mathbb{R}^{d_1} \to \mathbb{R}^1$ of class $C^{1,2}$ the Ito formula (1.2) is:

$$F(t, X_t) = F(0, X_0) + \int_0^t \left[\frac{\partial F}{\partial t}(s, X_s) + \sum_{i=1}^d b_s^i \frac{\partial F}{\partial x_i}(s, X_s)\right.$$

$$\left. + \frac{1}{2}\sum_{i=1}^d \sum_{j=1}^{d_1} \frac{\partial^2 F}{\partial x_i \partial x_j}(s, X_s) \sum_{k=1}^d \sigma_s^{ik}\sigma_s^{jk}\right] ds$$

$$+ \int_0^t \sum_{i=1}^d \sum_{j=1}^{d_1} \frac{\partial F}{\partial x_i}(s, X_s) \sigma_s^{ij} \sum_{k=1}^d \sigma_s^{jk} dW_s^j$$

The structure of diffusion processes as well as Ito's processes is determined by their drift and diffusion. The idea of singling out in the structure a low-frequency component (a process of bounded variation) and a high-frequency component (a martingale) leads to a much wider class of stochastic processes called semimartingales.

1.2 Contemporary stochastic analysis

Corresponding to the general theory of stochastic processes, we introduce, in addition to the stochastic basis $(\Omega, \mathcal{F}, \mathbb{F} = (\mathcal{F}_t)_{t \geqslant 0}, \mathrm{P})$, the following two σ-algebras on $\mathbb{R}_+ \times \Omega$:

- The σ-algebra $\mathcal{P}$ generated by all the $\mathbb{F}$-adapted continuous processes;

- The σ-algebra $\mathcal{O}$ generated by all the $\mathbb{F}$-adapted right-continuous processes.

These are the predictable and optional σ-algebras, and the corresponding $\mathcal{P}$-measurable and $\mathcal{O}$-measurable processes are called predictable processes and optional processes. One can give another and more constructive description of these σ-algebras using the notion of stopping times. A random variable $\tau : \Omega \to [0, \infty]$ is called a stopping time (Markov moment) if for every $t \geq 0$: $\{\omega : \tau(\omega) \leq t\} \in \mathcal{F}_t$. Any subset of $\mathbb{R}_+ \times \Omega$ is called a random set. Important examples of such sets are the so-called stochastic intervals: for the stopping times τ and σ,

$$[\![\sigma, \tau[\![= \{(t, \omega) : \sigma(\omega) \leq t < \tau(\omega) < \infty\},$$
$$[\![\sigma, \tau]\!] = \{(t, \omega) : \sigma(\omega) \leq t \leq \tau(\omega) < \infty\},$$

etc., and also their graphs $[\![\tau]\!]$, $[\![\sigma]\!]$.

For $A \in \mathcal{F}_0$, we can define a stopping time

$$O_A(\omega) = \begin{cases} 0, & \text{if } \omega \in A, \\ \infty, & \text{if } \omega \notin A. \end{cases}$$

The σ-algebras $\mathcal{P}$ and $\mathcal{O}$ admit the following description in terms of stochastic intervals:

$$\mathcal{P} = \sigma\{[\![O_A]\!] \text{ and } [\![o, \tau]\!] : A \in \mathcal{F}_0, \tau \text{ is a stopping time}\},$$

$$\mathcal{O} = \sigma\{[\![o, \tau[\![: \tau \text{ is a stopping time}\}.$$

A stopping time τ is called predictable if its graph is a predictable set. In this case one can find a non-decreasing sequence of stopping times $(\tau_n)_{n \geq 1}$ approximating τ in the following sense:

$$\tau(\omega) = \lim_{n \to \infty} \tau_n(\omega),$$

$$\tau_n(\omega) < \tau(\omega), \text{ on } \{\omega : \tau(\omega) > 0\}.$$

For a predictable process $X = (X_t)_{t \geq 0}$, its moments of jumps $\Delta X_t = X_t - X_{t-}$ are exhausted by a sequence of predictable stopping times $(\tau_n^X)_{n \geq 1}$ with non-intersecting graphs:

$$\{(t, \omega) : \Delta X_t(\omega) \neq 0\} = \bigcup_{n \geq 1} [\![\tau_n^X]\!].$$

A process $X = (X_t)_{t \geq 0}$ is called left quasi-continuous if it does not charge the predictable moments (for any predictable stopping time $\sigma : \Delta X_\sigma = 0$ (a.s.)).

A stopping time σ is called totally in accessible if for any predictable stopping time τ:

$$P\{\omega : \sigma(\omega) = \tau(\omega) < \infty\} = 0.$$

Thus, the moments of jumps of a left quasi-continuous process $X = (X_t)_{t \geq 0}$ are exhausted by a sequence of totally inaccessible stopping times in the sense given above for predictable processes.

A martingale $M = (M_t)_{t \geq 0}$ is said to be uniformly integrable if

$$\lim_{c \to \infty} \sup_{t \geq 0} \mathrm{E}\,|M_t|\,\mathbb{1}_{\{\omega:|M_t|>c\}} = 0.$$

The class of uniformly integrable martingales starting from zero is denoted by $\mathcal{M}$. For any $M \in \mathcal{M}$, there exists an integrable random variable M_∞ such that $M_\infty = \lim_{t \to \infty} M_t$ a.s. and $\mathrm{E}\,|M_t - M_\infty| \underset{t \to \infty}{\to} 0$, and $\mathrm{E}\,(M_\infty|\,F_t) = M_t$. A martingale $M = (M_t)_{t \geq 0}$ is square integrable ($M \in \mathcal{M}^2$) if $\sup_t \mathrm{E}|M_t|^2 < \infty$. Further, the set of processes having a finite (integrable) variation on each finite interval is denoted by $\mathcal{V}(\mathcal{A})$, and the subset of non-decreasing processes by $\mathcal{V}^+(\mathcal{A}^+)$.

A process $M = (M_t)_{t \geq 0}$ is a local martingale ($M \in \mathcal{M}_{loc}$) if there is a sequence of stopping times $\tau_n \uparrow \infty$ $(n \to \infty)$ such that $M^{\tau_n} = (M_{t \wedge \tau_n})$ belongs to $\mathcal{M}$ for each $n = 1, 2, \ldots$. The classes $\mathcal{M}^2_{loc}$, $\mathcal{A}^+_{loc}$, $\mathcal{A}_{loc}$ are defined similarly according to the indicated localization procedure, and the corresponding processes are called locally square-integrable martingales, and so on.

For a process $A \in \mathcal{V}^+ \equiv \mathcal{V}^+_{loc}$, we have a decomposition into the purely continuous component A^c and the purely discontinuous component A^d:

$$A_t = A_t^c + A_t^d,$$

where $A_t^d = \sum_{n \geq 1} d_n \mathbb{1}_{\{\omega:\tau_n(\omega) \leq t\}}$, and the sequence $(\tau_n)_{n \geq 1}$ of stopping times exhausts the moments of jumps of the process A.

If in the definition of martingales, the equality is replaced by the inequalities $\mathrm{E}\,(M_t|\,F_s) \geq M_s$ (a.s.) and $\mathrm{E}\,(M_t|\,F_s) \leq M_s$ (a.s.), then the process $M = (M_t, F_t)_{t \geq 0}$ is called a submartingale, and a supermartingale respectively. It turns out that the indicated processes are formed from local martingales (martingales) and, increasing predictable processes. For instance, every non-negative the supermartingale M admits the unique Doob–Meyer decomposition: for $t \geq 0$,

$$M_t = m_t - A_t,$$

where $m \in \mathcal{M}_{loc}$, $A \in \mathcal{A}^+_{loc} \cap \mathcal{P}$.

In particular, for $B \in \mathcal{A}^+_{loc}$, there exists a unique predictable process (compensator) $\tilde{B} \in \mathcal{A}^+_{loc} \cap \mathcal{P}$ such that $B - \tilde{B} \in \mathcal{M}_{loc}$.

The Doob–Meyer decomposition implies that any $M \in \mathcal{M}^2_{loc}$ has a quadratic characteristic $\langle M, M \rangle$, which is a predictable process in $\mathcal{A}^+_{loc}$ such that $M^2 - \langle M, M \rangle \in \mathcal{M}_{loc}$. For two such martingales $M, N \in \mathcal{M}^2_{loc}$ one can define their joint quadratic characteristic $\langle M, N \rangle \in \mathcal{A}^+_{loc} \cap \mathcal{P}$ satisfying the property $MN - \langle M, N \rangle \in \mathcal{M}_{loc}$. An arbitrary $M \in \mathcal{M}_{loc}$ does not admit such a characterization. Nevertheless, the following important decomposition into a continuous component M^c and a purely discontinuous component M^d holds true:

$$M_t = M_t^c + M_t^d,$$

where M^c is a continuous local martingale ($M^c \in \mathcal{M}^c_{loc}$), and M^d belong to the class $\mathcal{M}^d_{loc}$ of purely discontinuous local martingales, which is determined by the condition that $KL \in \mathcal{M}_{loc}$ for any $K \in \mathcal{M}^c_{loc}$ and $L \in \mathcal{M}^d_{loc}$.

Further, for $M, N \in \mathcal{M}_{loc}$, we define their joint quadratic characteristic:

$$[M, N]_t = \langle M^c, N^c \rangle_t + \sum_{0 < s \leq t} \Delta M_s \Delta N_s.$$

In particular, for $M \equiv N \in \mathcal{M}_{loc}$,

$$[M, M] - \langle M, M \rangle \in \mathcal{M}_{loc}.$$

Assume that $M \in \mathcal{M}^2_{loc}$ and the predictable process φ are such that the Lebesgue–Stieltjes integral (for almost all ω)

$$\varphi^2 \circ \langle M, M \rangle_t = \int_0^t \varphi_s^2 d\langle M, M \rangle_s$$

is defined and is in $\mathcal{A}^+_{loc}$. Then the stochastic integral

$$\varphi * M_t = \int_0^t \varphi_s dM_s$$

is well defined as a process in $\mathcal{M}^2_{loc}$ such that (a.s.)

$$\Delta(\varphi * M_t) = \varphi_t \Delta M_t \text{ and } \langle \varphi * M, \varphi * M \rangle_t = \varphi^2 \circ \langle M, M \rangle_t.$$

In the class $\mathcal{M}^2_{loc}$, all possible stochastic integrals with respect to a given $M \in \mathcal{M}^2_{loc}$ form a subspace, Consequently, any other $N \in \mathcal{M}^2_{loc}$ must have a decomposition into a sum of two projections (on the indicated subspace and its complement). This idea leads to the Kunita–Watanabe decomposition:

$$N_t = M'_t + \int_0^t \varphi_s dM_s,$$

where φ is a predictable process, and $M' \in \mathcal{M}^2_{loc}$ is orthogonal to M in the sense that $\langle M, M' \rangle = 0$.

A semimartingale is defined to be a process representable in the form

$$X_t = X_0 + A_t + M_t, \qquad (1.5)$$

where $A \in \mathcal{V}$, $M \in \mathcal{M}_{loc}$, and X_0 – $\mathcal{F}_0$-measurable random variable.

From the jumps of a d-dimensional semimartingale $X = (X_t)_{t \geq 0}$ in (1.5), we define the random measure

$$\mu(\omega, B, r) = \sum_{0 < s < \infty} \mathbb{1}_{\{(s, X_s(\omega)) \in (B, r)\}}, \qquad (1.6)$$

where $B \in \mathcal{B}(\mathbb{R}_+)$ and $r \in \mathcal{B}(\mathbb{R}_0^d)$, $\mathbb{R}_0^d = \mathbb{R}^d \setminus \{0\}$. The (integer-valued) random measure $\nu = \nu(\omega, B, r)$ is said to be predictable if for any $\mathcal{P} \times \mathcal{B}(\mathbb{R}_0^d)$-measurable non-negative function φ the process

$$\varphi \circ \nu_t = \int_0^t \int_{\mathbb{R}_0^d} \varphi(s,x) \nu(ds, dx)$$

is predictable. If in this case $E\varphi \circ \nu_\infty = E\varphi \circ \mu_\infty$, then ν is called the compensator μ defined by (1.6).

For a $\mathcal{P} \times \mathcal{B}(\mathbb{R}_0^d)$-measurable function φ such that $(|\varphi| \circ \nu_t) \in \mathcal{A}_{loc}^+$, one can define the stochastic integral with respect to the difference $(\mu - \nu)$ in measures as a purely discontinuous local martingale

$$\varphi * (\mu - \nu)_t = \int_0^t \int_{\mathbb{R}_0^d} \varphi(s,x) d(\mu - \nu), \qquad (1.7)$$

whose jumps are determined from the relation

$$\Delta \varphi * (\mu - \nu)_t = \varphi(t, \Delta X_t) \mathbb{1}_{\{\Delta X_t \neq 0\}} - \int_{\mathbb{R}_0^d} \varphi(t,x) \nu(\{t\}, dx).$$

Using (1.6)–(1.7), we can write the following canonical decomposition of X:

$$X_t = X_0 + B_t + M_t^c + x \cdot \mathbb{1}_{\{x:|x| \leq 1\}} * (\mu - \nu)_t + x \cdot I_{\{x:|x| > 1\}} \circ \mu_t, \qquad (1.8)$$

where the process $B \in \mathcal{V}$ is predictable, and $M^c \in \mathcal{M}_{loc}^c$.

The triple $(B, \langle M^c, M^c \rangle, \nu)$ is called the triplet of predictable characteristics of the semimartingale X.

It is now easy to define the stochastic integral with respect to a semimartingale X as a sum of the corresponding integrals with respect to the components of the decompositions (1.6) or (1.8). The class of semimartingales, like the classes of the Ito processes and diffusion processes is invariant with respect to smooth changes in variables. This is what is said by the corresponding generalization of the Ito formula given by Doleans and Meyer:

Let X be a semimartingale (1.6) with values in $\mathbb{R}^d$, $d \geq 1$, and let $F: \mathbb{R}^d \to \mathbb{R}^1$ be a twice continuously differentiable function. Then for all $t \geq 0$ (a.s.)

$$F(X_t) = F(X_0) + \int_0^t \sum_{i=1}^d \frac{\partial F}{\partial x_i}(X_{s-}) dX_s^i$$

$$+ \frac{1}{2} \int_0^t \sum_{i,j=1}^d \frac{\partial^2 F}{\partial x_i \partial x_j}(X_{s-}) d\langle X^{c,i}, X^{c,j} \rangle_s \qquad (1.9)$$

$$+ \sum_{0 < s \leq t} \left[F(X_s) - F(X_{s-}) - \sum_{i=1}^d \frac{\partial F}{\partial x_i}(X_{s-}) \Delta X_s^i \right],$$

where $X^c = M^c$ is the continuous part of a local martingale M.

The existence of stochastic integrals with respect to semimartingales enables us to consider corresponding stochastic integral (differential) equations. Of crucial significance for us here are linear stochastic equations

$$Y_t = Y_0 + \int_0^t Y_{s-} dX_s, \qquad (1.10)$$

where X is a given semimartingale (1.6).

A solution Y of equation (1.10) is determined with the help of the stochastic exponential

$$\varepsilon_t(X) = e^{X_t - X_0 - \frac{1}{2}\langle X^c, X^c \rangle_t} \prod_{0 < s \leqslant t} (1 + \Delta X_s) e^{-\Delta X_s} \qquad (1.11)$$

as follows $Y_t = Y_0 \varepsilon_t(X)$.

Let us list the most important properties of stochastic exponentials (1.11):

- $\varepsilon_t(X)$ is a process of bounded variation or a local martingale if X is;

- $\varepsilon_t^{-1}(X) = \varepsilon_t(-X^*)$, where $X_t^* = X_t - \langle X^c, X^c \rangle_t - \sum_{s \leqslant t} \frac{(\Delta X_s)^2}{1 + \Delta X_s}$, $\Delta X_s \neq -1$;

- $\varepsilon_t(X) \varepsilon_t\left(X'\right) = \varepsilon_t\left(X + X' + \left[X, X'\right]\right)$ (the multiplication rule for stochastic exponentials).

Suppose that besides the original measure on the stochastic basis $(\Omega, \mathcal{F}, \mathbb{F}, P)$, we are given a measure $\widetilde{P}$ locally equivalent to P and with local density $(Z_t)_{t \geqslant 0}$. The equivalence implies strict positivity of Z_t (P-a.s.) for $t \geqslant 0$. Hence, we can define a local martingale $N = (N_t)_{t \geqslant 0}$ with respect to P ($N \in \mathcal{M}_{loc}(P)$) as the stochastic integral $N_t = \int_0^t Z_{s-}^{-1} dZ_s$, which leads to a representation of Z_t in the form of the stochastic exponential $Z_t = \varepsilon_t(N)$. Just as a diffusion process acquires the properties of a Wiener process under a transformation of measure with the help of the Girsanov exponential, a given semimartingale R here acquires the properties of a local martingale with respect to a measure $\widetilde{P}$ with density Z_t expressed in the form of a stochastic exponential:

$$R \in \mathcal{M}_{loc}\left(\widetilde{P}\right) \Leftrightarrow ZR \in \mathcal{M}_{loc}(P).$$

There is another construction of (locally) equivalent measure $\widetilde{P}$. Let us describe it for a very important subclass of semimartingales called the Levy processes. The real-valued process $X = (X_t)_{t \geqslant 0}$, $X_0 = 0$, is the Levy process if

- $X_t - X_s$ is independent of $\mathcal{F}_s^X = \sigma\{X_u, u \leqslant s\}$ for all $s \leqslant t < \infty$;

- $X_{t+s} - X_s$ and X_s have the same distribution for all $s, t \geq 0$;
- X is continuous in probability, i.e., $\mathrm{P}\{\omega : |X_t - X_s| > \varepsilon\} \to 0$ when $s \to t$ for every $\varepsilon > 0$.

By the way, the last property means that the process X must be left quasi-continuous.

For a set $A \in \mathcal{B}(\mathbb{R}_0)$ and $t > 0$, we define

$$N_t^A(\omega) = \sum_{0 < s \leq t} \mathbb{1}_A(\Delta X_s(\omega)),$$

and the expected value $\mathrm{E} N_1^A = \nu(A)$ is said to be the Levy measure of X. The process $(N_t^A)_{t \geq 0}$ is a Poisson process with intensity $\nu(A)$ if A is fixed. One can derive the following canonical decomposition of the Levy process X as a semimartingale

$$X_t = bt + \sigma W_t + \int_0^t \int_{|x| \leq 1} x\left(N_t(dx) - t\nu(dx)\right) + \int_0^t \int_{|x| > 1} x N_t(dx),$$

where b and σ are constants, and W is a Wiener process.

If $\mathrm{E} e^{\alpha X_t} < \infty$, $\alpha > 0$, then the process $L_t = \frac{e^{\alpha X_t}}{\mathrm{E} e^{\alpha X_t}}$, $L_0 = 1$, is a positive martingale. This martingale generates a (locally) equivalent measure $\widetilde{P}$ with the (local) density $\frac{d\widetilde{P}_t}{dP_t} = L_t$. Such construction of $\widetilde{P}$ is known as the Esscher transform in mathematical finance and actuarial science.

For what follows, facts as the representation of any martingale in the form of a stochastic integral with respect to some basic martingale and the Doob–Meyer decomposition with respect to some family of measures will be important. For a given process $X = (X_t)_{t \geq 0}$ we denote $\mathcal{M}(X, \mathrm{P})$ the set of all measures $\widetilde{P} \sim \mathrm{P}$ such that X is a local martingale with respect to $\widetilde{P}$. This set will be called the family of martingale measures. The first key fact in this direction concerns the structure of nonnegative processes that are local martingales with respect to all $\widetilde{P} \in \mathcal{M}(X, \mathrm{P})$:

If $\mathcal{M}(X, \mathrm{P}) \neq \emptyset$, then a nonnegative process $M \in \mathcal{M}_{loc}(\widetilde{P})$ for any $\widetilde{P} \in \mathcal{M}(X, \mathrm{P}) \Leftrightarrow$ there exists a predictable process φ such that for any t a.s.

$$M_t = M_0 + \int_0^t \varphi_s dX_s. \tag{1.12}$$

The second key fact gives a description of the structure of nonnegative processes that are supermartingale with respect to all $\widetilde{P} \in \mathcal{M}(X, \mathrm{P})$ (optional decomposition or the uniform Doob–Meyer decomposition): Suppose $\mathcal{M}(X, \mathrm{P}) \neq \emptyset$ and V is a nonnegative process. Then V is a super-martingale for any $\widetilde{P} \in \mathcal{M}(X, P) \Leftrightarrow$ there exists a predictable process φ and a nonnegative optional process C such that for all $t \geq 0$ (a.s.)

$$V_t = V_0 + \int_0^t \varphi_s dX_s - C_t. \tag{1.13}$$

1.3 Mathematical finance and insurance

On a given stochastic basis $(\Omega, \mathcal{F}, \mathbb{F}, P)$, we define a (B, S)-market as a collection of two positive semimartingales B and S. The values of B_t and S_t are interpreted as the prices of a non-risky asset B and a risky asset S. A pair $\pi_t = (\beta_t, \gamma_t)$ of stochastic processes, with predictable second components is defined as a portfolio or investment strategy. The portfolio has a capital (value, wealth)

$$X_t^\pi = X_t^\pi(x) = \beta_t B_t + \gamma_t S_t, \tag{1.14}$$

where $X_0^\pi = \beta_0 B_0 + \gamma_0 S_0 = x$.

We assume for simplicity that the processes B and S are one-dimensional. All results below remain true for a d-dimensional asset $S = (S^1, ..., S^d)$ and a d-dimensional strategy $\gamma = (\gamma^1, ..., \gamma^d)$ if the product $\gamma_t S_t = \sum_{i=1}^d \gamma_t^i S_t^i$.

The ratio $X_t = \frac{S_t}{B_t}$ is called the discounted price, and $\frac{X_t^\pi}{B_t}$ is called the discounted capital of the strategy π.

Let us consider $\beta, \gamma \in \nu$ for simplicity and find from (1.14) with the help of the Ito formula

$$\frac{X_t^\pi}{B_t} = \frac{X_0^\pi}{B_0} + \int_0^t d\beta_u + \int_0^t \frac{S_{u-}}{B_{u-}} d\gamma_u + \int_0^t \gamma_u d\frac{S_{u-}}{B_{u-}}. \tag{1.15}$$

The last term in the equality (1.15) has a clear economic meaning as cumulative gains/losses of a portfolio from the stock change on the interval $[0, t]$. Hence, it is reasonable to distinguish those portfolios π such that

$$\frac{X_t^\pi}{B_t} = \frac{X_0^\pi}{B_0} + \int_0^t \gamma_u d\frac{S_u}{B_u}. \tag{1.16}$$

We call them self-financing and denote their collection SF.

We say that (B, S)-market admits an arbitrage at time $T > 0$ if there exits $\widetilde{\pi} \in SF$ such that (P-a.s.) $X_0^{\widetilde{\pi}} = 0$, $X_T^{\widetilde{\pi}} \geq 0$, and $P\left(\omega : X_T^{\widetilde{\pi}}(\omega) > 0\right) > 0$. In this case, the strategy $\widetilde{\pi}$ is called the arbitrage strategy. It is well known that for many financial markets $((B, S)$-market), the absence of arbitrage can be characterized in terms of martingale measures as follows (First Fundamental Theorem): $\mathcal{M}(X, \mathrm{P}) \neq \varphi$, where $X = \frac{S}{B}$.

Let us extend the set SF defined by (1.16) forming the bigger class of strategies π with consumption C as a nonnegative non-decreasing process for which

$$X_t^\pi = X_0^\pi + \int_0^t \beta_u dB_u + \int_0^t \gamma_u dS_u - C_t \qquad (1.17)$$

or, equivalently,

$$\frac{X_t^\pi}{B_t} = \frac{X_0^\pi}{B_0} + \int_0^t \gamma_u dX_u - D_t, \qquad (1.18)$$

where $D_t = \int_0^t B_u^{-1} dC_u$.

Now we are ready to provide a natural interpretation of the sum of the second and the third terms in (1.15) as the cumulative (discounted) consumption.

Both induced classes of strategies admit the following martingale characterization:

If $\mathcal{M}(X, \mathrm{P})$ consists of a simple measure P* and Y is a nonnegative process, then:

- Y is the discounted value of a self-financing strategy $\Leftrightarrow$ Y is a local martingale with respect to P*;

- Y is the discounted value of a strategy with consumption $\Leftrightarrow$ Y is a (local) supermartingale with respect to P*.

In the first case, the proof follows from the martingale representation theorem (1.12), and in the second from the Doob–Meyer decomposition or its generalization (1.13). By a contingent claim with exercise time T, we understand any nonnegative $\mathcal{F}_T$-measurable random variable f. We remark that if such claim f represents a pay-off of an option, then the option is called a European option. A strategy $\pi \in SF$ is said to be a hedge (hedging strategy, perfect hedge) for f (or for the option with the pay-off function f) if $X_T^\pi(x) \geq f$ (a.s.) for some initial capital x.

If in some class of hedging strategies π, there is a strategy π^* such that $X_t^{\pi^*} \leq X_t^\pi$ (a.s.) for all $t \in [0, T]$, then π^* is called a minimal hedge (in this class). Very often, the minimal hedge coincides with a replicating strategy π^* for which $X_t^{\pi^*} = f$ (a.s.). In this case, f is called attainable. The (B, S)-market is said to be complete if any contingent claim is attainable. For several financial markets, this notion is equivalent to the uniqueness of P*, i.e., $\mathcal{M}(X, \mathrm{P}) = \{\mathrm{P}^*\}$.

Such characterization of complete markets is usually called the second fundamental theorem of financial mathematics. Under the uniqueness of a martingale measure P* (complete market) and the assumption $E^*\left(\frac{f}{B_T}\right) < \infty$ one can prove with the help of martingale representation the existence and uniqueness of a minimal hedge $\pi_t^* = (\beta_t^*, \gamma_t^*) \in SF$ such that

$$\frac{X_t^{\pi^*}}{B_t} = E^*\left(\frac{f}{B_T}\bigg| F_t\right) = \frac{X_0^{\pi^*}}{B_0} + \int_0^t \gamma_u^* dX_u, \text{ (a.s.)}, \ t \in [0, T]. \quad (1.19)$$

It follows from (1.19) that the natural (fair) price for the option with payoff f should be $E^*\left(\frac{f}{B_T}\right)$. This is a methodological base for option pricing in complete markets.

In case of incomplete markets, where the set $\mathcal{M}(X, P)$ is not single, pricing and hedging problems can be solved in a class of strategies with consumption (see (1.17)–(1.18)). Namely, for given contingent claim f satisfying the condition $\sup_{\tilde{P} \in \mathcal{M}(X,P)} \tilde{E}\left(\frac{f}{B_T}\right) < \infty$, there exists a unique minimal hedge π^* with discounted consumption D^* such that

$$\frac{X_t^{\pi^*}}{B_t} = \operatorname*{ess\,sup}_{\tilde{P} \in \mathcal{M}(X,P)} \tilde{E}\left(\frac{f}{B_T}\bigg| F_t\right) = \frac{X_0^{\pi^*}}{B_0} + \int_0^t \gamma_u^* dX_u - D_t^*. \quad (1.20)$$

The proof is based on the optional decomposition. It follows from (1.20) that the natural (upper) initial price of option f will be equal to $\operatorname*{ess\,sup}_{\tilde{P} \in \mathcal{M}(X,P)} \tilde{E}\left(\frac{f}{B_T}\right)$, $B_0 = 1$. These results present a natural methodology of (super) hedging and pricing of contingent claims in incomplete markets.

Summing up these results, one can speak of the necessity not only of establishing the nonemptiness itself of the class $\mathcal{M}(X, P)$, but also of being able to find martingale measures. That is why we present here a general method for finding such measures in a very general (B, S)-market model whose prices B_t and S_t satisfy linear stochastic differential equations with respect to semimartingales. Most financial markets are modeled in just this way, and hence, the results for different market models can be derived as natural corollaries.

Suppose that a (B, S)-market is determined by the two equations

$$B_t = B_0 + \int_0^t B_{u-} dh_u, \ \Delta h_u > -1,$$

$$S_t = S_0 + \int_0^t S_{u-} dH_u, \ \Delta H_u > -1, \quad (1.21)$$

or

$$B_t = B_0 \varepsilon_t(h) \text{ and } S_t = S_0 \varepsilon_t(H),$$

where h and H are given semimartingales.

The problem here is to determine conditions under which a measure P* equivalent to P takes the process $X = \frac{S}{B}$ into a local martingale, i.e.,

$$P^* \in \mathcal{M}(X, P) \Leftrightarrow X = \frac{S}{B} \in \mathcal{M}_{loc}(P^*). \quad (1.22)$$

Let us investigate (1.22) for the market (1.21) when the initial measure P is a martingale measure. According to the properties of stochastic exponentials we obtain

$$X_t = X_0 \varepsilon_t(H) \varepsilon_t^{-1}(h) = X_0 \varepsilon_t(H) \varepsilon_t(-h^*)$$

$$= X_0 \varepsilon_t \left(H - h + \langle h^c, h^c \rangle + \sum \frac{(\Delta h)^2}{1 + \Delta h} - \langle H^c, h^c \rangle - \sum \frac{\Delta H \Delta h}{1 + \Delta h} \right) \quad (1.23)$$

$$= X_0 \varepsilon_t \left(H - h + \langle h^c, h^c - H^c \rangle + \sum \frac{\Delta h (\Delta h - \Delta H)}{1 + \Delta h} \right)$$

Let us denote

$$\Psi_t(h, H) = H_t - h_t + \langle h^c, h^c - H^c \rangle_t + \sum_{s \leq t} \frac{\Delta h_s (\Delta h_s - \Delta H_s)}{1 + \Delta h_s}$$

and rewrite (1.23) as

$$X_t = X_0 + \int_0^t X_{s-} d\Psi_s(h, H).$$

Consequently, to verify that $X \in \mathcal{M}_{loc}(P)$, we must establish that $\Psi(h, H) \in \mathcal{M}_{loc}(P)$. It is clear how to find desirable conditions for the relation (1.22). To this end, let $Z_t^* = \frac{dP_t^*}{dP_t}$ be the local density of P* and P. Then (as in Paragraph 1.2)

$$X \in \mathcal{M}_{loc}(P^*) \Leftrightarrow ZX \in \mathcal{M}_{loc}(P) \quad (1.24)$$

and

$$Z_t = \varepsilon_t(N), \quad (1.25)$$

where $N_t = \int_0^t Z_{s-}^{-1} dZ_s \in \mathcal{M}_{loc}(P)$.

Hence, we can transform (1.24) with the help of (1.25) as follows

$$X \in \mathcal{M}_{loc}(P^*) \Leftrightarrow X\varepsilon(N) \in \mathcal{M}_{loc}(P). \quad (1.26)$$

In view of (1.23)

$$X_t \varepsilon_t(N) = X_0 \varepsilon_t(\Psi(h, H)) \varepsilon_t(N),$$

and by the multiplicative rule for stochastic exponentials,
$$X_t \varepsilon_t(N) = X_0 \varepsilon_t(\Psi(h, H, N)), \qquad (1.27)$$
where
$$\Psi_t(h, H, N) = H_t - h_t + N_t + \langle (h - N)^c, (h - H)^c \rangle_t$$
$$+ \sum_{s \leqslant t} \frac{(\Delta h_s - \Delta N_s)(\Delta h_s - \Delta H_s)}{1 + \Delta h_s}.$$

It follows from (1.27) that to verify (1.26), we must determine when
$$\Psi(h, H, N) \in \mathcal{M}_{loc}(P). \qquad (1.28)$$

Let us see how this methodology works for some well-known models of a (B, S)-market. The Black–Scholes model (written in a differential form):
$$\begin{aligned} dB_t &= rB_t dt, \ B_0 = 1 \\ S_t &= S_t(\mu dt + \sigma dW_t), \ S_0 > 1, \end{aligned} \qquad (1.29)$$

where $r, \mu \in \mathbb{R}^+$, the volatility $\sigma > 0$, and $W = (W_t)_{t \geqslant 0}$ is a Wiener process generating filtration $\mathbb{F}$.

In the model (1.29), $h_t = rt$ and $H_t = \mu dt + \sigma dW_t$, and since W is the sole source of randomness, it is natural to look for the local martingale N in the form $N_t = \varphi \cdot W_t$. Using this, we get from (1.27) that
$$\begin{aligned} \Psi_t(h, H, N) &= \mu dt + \sigma dW_t - rt + \varphi \cdot W_t + \varphi \sigma t \\ &= (\mu - r + \varphi \sigma) t + (\sigma + \varphi) W_t \end{aligned}$$

Therefore, the condition (1.28) will be satisfied if $\mu - r + \varphi \sigma = 0$ or $\varphi = -\frac{\mu - r}{\sigma}$. Using (1.25), we can construct the (local) density
$$Z_t^* = \frac{dP_t^*}{dP_t} = \varepsilon_t(N) = \exp\left\{ -\frac{\mu - r}{\sigma} W_t - \frac{1}{2}\left(\frac{\mu - r}{\sigma}\right)^2 t \right\}. \qquad (1.30)$$

Moreover, the process $W_t^* = W_t + \frac{\mu - r}{\sigma} t$ will be a Wiener process by the Girsanov theorem with respect to the unique martingale measure P^*. Let us consider the contingent claim $f = (S_T - K)^+$ corresponding to a call option, the option to buy the asset S at the (strike) price K. According to the methodology of pricing and hedging contingent claims in completed markets, there exists a minimal hedge $\pi_t^* = (\beta_t^*, \gamma_t^*)$ for such an option. We have, using (1.30) and omitting standard technical details
$$\begin{aligned} X_t^{\pi^*} &= E^*\left(e^{-r(T-t)} (S_T - K)^+ \big| \mathcal{F}_t \right) \\ &= S_t \Phi(d_+(T-t)) - K e^{-r(T-t)} \Phi(d_-(T-t)) \\ &= \mathbb{C}^{BS}(T-t, S_t, K, r, \sigma) = \mathbb{C}^{BS}(t, S_t), \end{aligned} \qquad (1.31)$$

where $d_\pm(T-t) = \frac{\ln\frac{S_t}{K} + (T-t)\left(r \pm \frac{\sigma^2}{2}\right)}{\sigma\sqrt{T-t}}$.

The formula (1.31) is called the Black–Scholes formula. Obviously, the option price presented by (1.31) is a smooth function with respect to time and space variables. Therefore, applying to $\mathbb{C}^{BS}(t, S_t)$ the Ito formula, we get

$$\frac{\mathbb{C}^{BS}(t, S_t)}{B_t} = \frac{\mathbb{C}^{BS}(0, S_0)}{B_0} + \int_0^t \frac{\partial \mathbb{C}^{BS}}{\partial x}(u, S_u) \, d\left(\frac{S_u}{B_u}\right)$$

$$+ \int_0^t B_u^{-1} \left[\frac{\partial \mathbb{C}^{BS}}{\partial u}(u, S_u) + rS_u \frac{\partial \mathbb{C}^{BS}}{\partial x}(u, S_u) \right. \quad (1.32)$$

$$\left. + \frac{\sigma^2}{2} \frac{\partial^2 \mathbb{C}^{BS}}{\partial^2 x}(u, S_u) - r\mathbb{C}^{BS}\right] du.$$

Using properties of the capital of minimal hedge, we obtain from (1.32)

$$\gamma_t^* = \frac{\partial \mathbb{C}^{BS}}{\partial x}(t, S_t), \quad (1.33)$$

$$\frac{\partial \mathbb{C}^{BS}}{\partial t}(t, x) + rx\frac{\partial \mathbb{C}^{BS}}{\partial x}(t, x) + \frac{\sigma^2}{2}x^2\frac{\partial^2 \mathbb{C}^{BS}}{\partial^2 x}(t, x) - r\mathbb{C}^{BS}(t, x), \quad (1.34)$$

where $x = S_t$.

The formula (1.34) presents the famous Black–Scholes differential equation.

The second component β_t^* of the minimal hedge can be reconstructed with the help of (1.33) and the balance equation $X_t^{\pi^*} = \beta_t^* B_t + \gamma_t^* S_t$ as follows:

$$\beta_t^* = B_t^{-1}\left(X_t^{\pi^*} - \gamma_t^* S_t\right). \quad (1.35)$$

So, in the framework of the Black–Scholes model (1.29) the components of the minimal hedge $\pi_t^* = (\beta_t^*, \gamma_t^*)$ can be found from formulas (1.33) and (1.35), and its value is determined by formula (1.31) and satisfies equation (1.34).

As for the put option, the option to sell the stock S at price K, its price $\mathbb{P}^{BS}(T)$ can be calculated by the following call-put parity:

$$\mathbb{P}^{BS}(T) = \mathbb{C}^{BS}(T) - S_0 + Ke^{-rT}.$$

The jump-diffusion model of a market:

$$dB_t = rB_t dt, \; B_0 = 1$$
$$dS_t^i = S_{t-}^i \left(\mu^i dt + \sigma^i dW_t - \nu^i d\Pi_t\right), \; S_0^i > 1, \; i = 1, 2, \quad (1.36)$$

where $r, \mu^i \in \mathbb{R}^+$, $\sigma^i > 0$, $\nu^i < 1$, $W = (W_t)_{t\geq 0}$ is a Wiener process, $\Pi = (\Pi_t)_{t\geq 0}$ is a Poisson process with intensity $\lambda > 0$.

As in model (1.29), we calculate the function $\Psi_t(h, H, N)$, and then we look for the martingale $N = (N_t)_{t \geq 0}$ in the form $N_t = \varphi W_t + \psi (\Pi_t - \lambda t)$: for $i = 1, 2$,

$$\Psi_t(h, H, N) = \mu^i dt + \sigma^i dW_t - \nu^i \Pi_t - rt - \nu^i \psi \Pi_t + \varphi \sigma^i t$$
$$= \left(\mu^i - r - \nu^i \lambda + \varphi \sigma^i - \psi \nu^i \lambda\right) t + \text{(a martingale)}.$$

Hence, φ and ψ satisfy the two equations ($i = 1, 2$):

$$\mu^i - r - \nu^i \lambda + \varphi \sigma^i - \psi \nu^i \lambda = 0 \tag{1.37}$$

or, with the notation $\lambda^* = \lambda(1 + \psi)$,

$$\mu^i - r - \nu^i \lambda + \varphi \sigma^i - \nu^i \lambda^* = 0. \tag{1.38}$$

Assuming that $\sigma^2 \nu^1 - \sigma^1 \nu^2 \neq 0$, we can find the unique solution for (1.37)–(1.38):

$$\varphi = \frac{\left(\mu^2 - r\right) \nu^1 - \left(\mu^1 - r\right) \nu^2}{\sigma^2 \nu^1 - \sigma^1 \nu^2},$$

$$\lambda^* = \frac{\left(\mu^1 - r\right) \sigma^2 - \left(\mu^2 - r\right) \sigma^1}{\sigma^2 \nu^1 - \sigma^1 \nu^2}.$$

The density of the martingale measure P* can be uniquely determined from this:

$$Z_t^* = \frac{dP_t^*}{dP_t} = \varepsilon_t(N)$$
$$= \exp\left\{\varphi W_t - \frac{\varphi^2 t}{2} + \psi(\Pi_t - \lambda t)\right\} \prod_{s \leq t} (1 + \psi \Delta \Pi_s) e^{-\psi \Delta \Pi_s} \tag{1.39}$$
$$= \exp\left\{\varphi W_t - \frac{\varphi^2 t}{2} + (\lambda - \lambda^*) t + (\ln \lambda^* - \ln \lambda) \Pi_t\right\}.$$

We note that the process $W_t^* = W_t - \varphi t$ is a Wiener process, while Π_t is a Poisson process with intensity λ^* with respect to P* defined by (1.39).

We remark that the consideration of a (B, S)-market with a single stock leads only to the single equation (1.37) with two unknowns. In this case, the market admits (in general) many martingale measures, and hence, the market is incomplete.

In the framework of the model (1.36), we can derive the formula for the call price

$$\mathbb{C} = E^* e^{-rT} \left(S_T^1 - K\right)^+$$

on the first or second stock with the strike price K.

So, using (1.39) we can calculate (again, omitting the technical details):

$$\mathbb{C} = e^{-\lambda^* T} \sum_{n=0}^{\infty} \frac{(\lambda^* T)^n}{n!} \mathbb{C}^{BS}\left(T, S_0^1 (1 - \nu^1)^n e^{\nu^1 \lambda^* T}, r, \sigma^1, K\right)$$

as the Poisson weighing of the Black–Scholes formula.

Up to now, we operated only within the case where the so-called perfect hedging was possible (in class SF):

$$P\left\{\omega : X_T^{\pi^*}(\omega) = f\right\} = 1. \tag{1.40}$$

It was also noted that (1.40) is not possible (in general) in incomplete markets. Another very natural situation is the case of market constraints. Our attention below will be concentrated mainly on the so-called budget constraints of the initial capital of the portfolio:

$$X_0^{\pi} \leqslant \tilde{x} < E^*\left(\frac{f}{B_T}\right). \tag{1.41}$$

What kind of hedging can be proposed under this condition (1.41)?

Let us imagine a hedging procedure as a way to construct a strategy π^* such that its terminal capital is close enough to a given contingent claim f in some probabilistic sense. This viewpoint into hedging leads us to the following types of hedging:

- Mean-variance hedging, when $X_T^{\pi^*}$ is closest to f in L^2;

- Quantile hedging, when $X_T^{\pi^*}$ is closest to f in the sense of maximization of probability of successful hedging:

$$P\left\{\omega : X_T^{\pi^*} \geqslant f\right\} = \max_{\pi} P\left\{\omega : X_T^{\pi} \geqslant f\right\};$$

- Efficient hedging, when $X_T^{\pi^*}$ is most closed to f in the sense:

$$E\, l\left(X_T^{\pi^*} - f\right) = \min_{\pi} E\, l\left(X_T^{\pi} - f\right),$$

where $l = l(x)$, $x \in \mathbb{R}^+$, is a loss function.

The notion borrowed from mathematical statistics to estimate the quality of some statistical procedures.

Any derivative security (forwards, futures, options, etc.,) is characterized by some insurance properties. They are exploited to avoid losses or to minimize risks. Traditional insurance focuses exclusively on the estimation of risks. The circulation of derivative financial instruments is quite similar to insurance in many aspects. However, while insurance companies have to hold considerable reserves of the capital in relation to their liabilities, a more flexible option insurance reduces these requirements for reserves. The principal reason for this situation is apparently that traditional insurance involves client's selling the risk to a specific insurance company, while insurance on the basis of derivative securities involves selling the same risk on a financial market with the possibility of dynamic observation of prices, and an adequate reaction to price changes in basis assets of the market. So, the significance of the new instruments of insurance naturally connected to the risky assets of the financial markets was fully recognized. For instance, in life insurance, there appeared contracts in which the payment by the maturity date was connected with the market value of a portfolio of assets. These are the equity-linked life insurance contracts, that is, insurance linked to the risky assets in the financial market. The appearance in actuarial science and practice of the Black–Scholes formula so crucial for mathematical finance is therefore not surprising.

The contracts we study have two types of uncertainty:

- Financial market uncertainty (financial risk);

- Client mortality (insurance risk).

If $T(x)$ is the future life time of a client of age x, then $_tp_x = P(\omega : T(x) > t)$ is called a survival probability. And the pay-off function of a finance-insurance contract under consideration is a function of stock prices during the contract period $[0,T]$ and $T(x)$. The terminology used in this regard is the following:

- Equity (unit)linked life insurance (Europe);

- Variable annuities (USA);

- Segregated funds (Canada).

We will use the European terminology, which reflects the nature of these contracts in clearest form. From a variety of such contracts (death guarantees, maturity guarantees, resets, fund switching, etc.,) we concentrate our attention on maturity and death guarantees. The pay-off function of the maturity guarantee contract can be expressed in the form:

$$\max(K, S_T) \cdot \mathbb{1}_{\{T(x)>T\}}, \qquad (1.42)$$

where K is a maturity guarantee.

We will study the maturity guarantee contracts using partial/imperfect hedging. Let us explain why this approach and technique can be successfully converted to a real methodology for pricing, hedging and risk-management of such a contract.

In traditional insurance (without a financial market), the pricing of (1.42)) is split into two steps:

- Calculate the discounted value of guarantee K: $e^{-rT}K$;

- Find the survival probability $_Tp_x$ and finally obtain the price as a product:

$$_Tp_x \cdot e^{-rT}K. \qquad (1.43)$$

If the payoff (1.42) contains a stock price component (assume, for certainty, S_t follows the Black–Scholes model), then we note

$$\max(K, S_T) = K + \max(0, S_T - K) = K + (S_T - K)^+. \qquad (1.44)$$

The relation (1.44) shows that the initial contract contains two pay-off components K and $(S_T - K)^+$. The second component is called an embedded (call) option. So, applying the option pricing theory, we naturally arrive at the following formula for the price $\mathbb{C}$ of the contract (1.42):

$$\begin{aligned}\mathbb{C} &= {}_T p_x \cdot e^{-rT} K + {}_T p_x \mathbb{C}_T^{BS} \\ &= {}_T p_x \cdot e^{-rT} K + {}_T p_x \left(S_0 \Phi(d_+(T)) - e^{-rT} K \Phi(d_-(T)) \right).\end{aligned} \qquad (1.45)$$

We will call (1.45) the Brennan–Schwartz price of the contract with the payoff (1.42). The second component $\widetilde{\mathbb{C}}$ of (1.45) is the price of the embedded call option multiplied by ${}_T p_x$, and therefore, it is less than the Black–Scholes price for call option $\mathbb{C}_T^{BS}$:
$\widetilde{\mathbb{C}} < \mathbb{C}^{BS}$.

It means the amount $\widetilde{\mathbb{C}}$ is not enough for a perfect hedge for $(S_T - K)^+$, and, hence, the price $\mathbb{C}$ is not enough for a perfect hedge for $\max(K, S_T)$. These observations call for partial/imperfect hedging in this area, and the book systematically develops this fruitful idea for both different types of market and of hedging.

The survival probability ${}_T p_x$ presented in the formulas (1.43)) and (1.45) gives a quantitative value for the mortality. Practically all calculations in life insurance are carried out on the basis of the demographic history of the insurance events including death, sickness, etc. For these purposes, one uses mortality (life) tables giving a general idea about the probabilities of such events. Mortality tables are statistical data about the longevity of a population group, categorized according to different criteria (gender, region, profession, etc.). They contain a number of variables and indicators characterizing the decrease in numbers of the group under observation and the mortality level for different periods.

Chapter 2

Quantile hedging of equity-linked life insurance contracts in the Black–Scholes model

2.1	Quantile hedging of contracts with deterministic guarantees ...	23
2.2	Quantile hedging of contracts with stochastic guarantees	32
2.3	Quantile portfolio management with multiple contracts with a numerical example	46

2.1 Quantile hedging of contracts with deterministic guarantees

In the complete financial market, every contingent claim can be hedged perfectly, and in an incomplete market, it is possible to stay on the safe side by super-hedging. In many situations, super-hedging needs a large amount of initial capital to set up the portfolio, which seems costly from the capital point of view. In this case, the following questions naturally arise. What if the investor is unwilling or unable to put up the large amount of initial capital required for hedging or super-hedging? Are we able to construct a hedging strategy so that the investor can achieve the maximal probability of a successful hedge with a smaller amount of initial capital? The answers can be fand with the help of quantile hedging.

There exist two forms of problems involved with quantile hedging. The first one is to minimize the value of a minimal hedging given the hedging probability; the second problem is to maximize the hedging probability given a constraint on the initial value of a minimal hedge. In this sense, such a hedging problem is methodologically related to the problem of statistical confidence estimation. The quantile is one of the main concepts in the general theory of statistical estimation, which is the boundary of the domain of estimation with a specific probability. So, the approach of hedging with probability less than 1 has come to be named as "quantile hedging".

As an imperfect or partial hedging technique, quantile hedging is based on the fundamental statistical result called the Neyman–Pearson lemma. Suppose we want to test the null hypothesis H_0 with probability measure P_0, against the alternative hypothesis H_1 with probability measure P_1. A type I error α

is defined as rejecting H_0 when it is true, and a type II error β is defined as accepting H_0 when it is false. Note that $\alpha, \beta \in (0,1)$ are probabilities of corresponding events. Generally, the probability measure corresponding to the alternative hypothesis is considered the real-world probability measure or the objective measure. The aim of the test is to reject H_0 when it is indeed false. During the test of two hypotheses, we need to control the size of Type I error α while minimizing the Type II error β. Equivalently, we can fix α and maximize the power of the test $1 - \beta$. The Neyman–Pearson lemma is stated as follows.

Lemma 2.1. *If the test ϕ satisfies*

$$E_0(\phi(X)) = \alpha, \tag{2.1}$$

$$\phi = \begin{cases} 1 & \text{when } \frac{dP_1}{dP_0} > c; \\ 0 & \text{when } \frac{dP_1}{dP_0} < c \end{cases} \tag{2.2}$$

for some constant c, then it is the most powerful for testing measure P_0 against measure P_1 at level α.

The conclusion of the Neyman–Pearson lemma is the basis for our partial hedging analysis. It provides the structure of the set on which the power of the test $1 - \beta$ is maximized with a given significance level α.

Assume the risk-free interest rate is a constant r and the discounted price process $X = (X_t)_{t \in [0,T]}$ of the underlying risky asset is a (semimartingale) process on probability space $(\Omega, \mathcal{F}, P)$ with the filtration $\mathcal{F} = (\mathcal{F}_t)_{t \in [0,T]}$. Let $\mathbf{P}$ denote the set of all equivalent martingale measures. If the set $\mathbf{P}$ is non-empty, there is no arbitrage opportunity in the market.

A self-financing strategy $\pi = (V_0, \xi_t)$ is defined by an initial capital $V_0 \geqslant 0$ and a predictable process ξ_t so that the corresponding value process V_t satisfies:

$$V_t = V_0 + \int_0^T \xi_s dX_s, \ \forall t \in [0,T], \ P - \text{a.s.} \tag{2.3}$$

Such strategy π is admissible if

$$V_t \geqslant 0, \ \forall t \in [0,T], \ P - \text{a.s.} \tag{2.4}$$

We consider a contingent claim whose payoff H is $\mathcal{F}_T$-measurable. A successful hedging set A is defined as $A = \{\omega : V_T \geqslant He^{-rT}\}$. The completeness of the market implies the existence of a unique martingale measure $P^* \sim P$, such that payoff H can be hedged perfectly with the required initial cost $H_0 = E^*\left(He^{-rT}\right)$, where E^* denotes the expectation with respect to P^*. With the unique hedging strategy, we also have $P(A) = 1$. However, if the investor is unable to allocate the required initial capital for perfect hedging, what is the best hedge he can achieve with a smaller amount $\widetilde{V}_0 < H_0$?

The problem can be formulated as the problem to construct an admissible strategy V_0, ξ_t such that

$$\mathrm{P}\left(V_T = V_0 + \int_0^T \xi_s dX_s \geq H\right) \to \max_{(V_0, \xi)} \qquad (2.5)$$

with constraint

$$V_0 \leq \tilde{V}_0 < H_0. \qquad (2.6)$$

Lemma 2.2. *Let $A^* \in \mathcal{F}_T$ be the solution of the following problem:*

$$\mathrm{P}(A) \to \max, \qquad (2.7)$$

$$\mathrm{E}^*(He^{-rT}\mathbb{1}_A) \leq \tilde{V}_0 < H_0, \qquad (2.8)$$

where $\mathbb{1}_{\{\cdot\}}$ is the indicator function. Then a perfect hedge π with initial value V_0 for the contingent claim $H^* = H \cdot \mathbb{1}_A$ is the solution of the problem (2.5)–(2.6), and the successful hedging set A coincides with A^*.

Lemma 2.2 states that the constructed optimal hedge π, called a quantile hedge, is the perfect hedge for a modified contingent claim H^*. The payoff H can be hedged with maximal probability on the set A^*, which is also called the maximal successful hedging set. Based on the Neyman–Pearson lemma, the structure of A^* can be obtained. We can define the measure Q^* with density

$$\frac{d\mathrm{Q}^*}{d\mathrm{P}^*} = \frac{H}{H_0}. \qquad (2.9)$$

It is assumed that Q^* corresponds to measure P_0 and P corresponds to measure P_1 in Lemma 2.1. Under the measure Q^*, the budget constraint $\mathrm{E}^*\left(He^{-rT}\mathbb{1}_{A^*}\right) \leq \tilde{V}_0$ becomes

$$\mathrm{Q}^*(A) \leq \frac{\tilde{V}_0}{H_0} = \frac{\tilde{V}_0}{\mathrm{E}^*(He^{-rT})}. \qquad (2.10)$$

We maximize the power of the hypothesis test given the Type I error α equal to $\frac{\tilde{V}_0}{H_0}$. According to the Neyman–Pearson lemma, the maximal successful hedging set A^* has the following structure

$$A^* = \left\{\frac{d\mathrm{P}}{d\mathrm{P}^*} > a^* H e^{-rT}\right\}, \qquad (2.11)$$

where $\frac{d\mathrm{P}}{d\mathrm{P}^*}$ is the density of the equivalent martingale measure, and a^* is a constant determined from the condition $a^* = \inf\left\{a : \mathrm{Q}^*\left(\frac{d\mathrm{P}}{d\mathrm{Q}^*} > a\right) \leq \alpha\right\}$. Using the structure of the successful hedging set A^* we can calculate the explicit expectations for initial cost V_0 and for the number ξ of units of risky assets.

We work in the Black and Scholes setting. Namely, it is a financial market with a constant interest $r \geqslant 0$, one riskless money market account B and one risky asset S satisfying:

$$dB_t = rB_t dt, \ B_0 = 1, \tag{2.12}$$

$$dS_t = S_t(\mu td + \sigma dW_t), \tag{2.13}$$

where W is a Wiener process, $\mu \in \mathbb{R}$, $\sigma > 0$.

All processes are given on a standard stochastic basis $(\Omega, \mathcal{F}, (\mathcal{F}_t)_{t \geqslant 0}, P)$ and are adapted to the filtration $\mathcal{F}$, generated by $(W_t)_{t \geqslant 0}$. Every predictable process $\pi = (\pi_t)_{t \geqslant 0} = (\beta, \gamma)_{t \geqslant 0}$ is called a trading strategy with the value (capital)

$$X_t^\pi = \beta_t B_t + \gamma_t S_t. \tag{2.14}$$

Only self-financing (with no additional inflow/outflow of cash other than initial price payment) and admissible (with nonnegative capital) strategies are allowed.

Let a positive random variable $T(x)$ on a probability space $(\tilde{\Omega}, \tilde{\mathcal{F}}, \tilde{P})$ be the remaining lifetime of a person of current age x. Denote $_T p_x = \tilde{P}(T(x) > T)$ the survival probability of the corresponding policyholder, which means the probability of a life aged x to survive T more years. We make the standard assumption that the insurance risk arising from clients' mortality and the financial market risk have no effect on each other, or are independent in probabilistic terminology.

We consider a single-premium equity-linked life insurance contract, which entitles the client to one unit of a risky asset or a guaranteed amount whichever is greater at expiration date T. The payoff has the form

$$H = \max\{S_T, K_T\} = S_T \mathbb{1}_{\{S_T \geqslant K_T\}} + K_T \mathbb{1}_{\{S_T \leqslant K_T\}}, \tag{2.15}$$

where $K = K_T$ is the deterministic guarantee, calculated as $K_T = S_0 \exp(gT)$, and g is the rate guaranteed by the contract, and S_0 is the initial value of the risky fund. Basically, the client has the right to choose the larger of two funds at maturity of the contract: a risky fund with expected return μ or a risk-free fund earning a rate of g over the duration of the agreement. We know that in this setting, a unique risk neutral measure P* exists, and its density is given by

$$Z_t = \left.\frac{dP^*}{dP}\right|_{\mathcal{F}_t} = \exp\left(-\theta W_t - \frac{\theta^2}{2}t\right), \ \theta = \frac{\mu - r}{\sigma}. \tag{2.16}$$

The payoff H occurs on or after the maturity of the policy. Therefore, the policyholder receives H provided that he/she is alive to collect the playoff H^*:

$$H^* = H \cdot \mathbb{1}_{T(x) > T}, \tag{2.17}$$

where $\mathbb{1}$ is the indicator function that equals one if the insured survives to time T or zero if the insured dies before T.

The fair premium U_0 for this policy is the so-called Brennan–Schwartz price

$$U_0 = \mathrm{E}^* \times \widetilde{\mathrm{E}} \left(H e^{-rT} \mathbb{1}_{\{T(x) > T\}} \right)$$
$$= \mathrm{E}^* \left(H e^{-rT} \right) \widetilde{\mathrm{P}} \{T(x) > T\} = \mathrm{E}^* \left(H e^{-rT} \right) \cdot {}_T p_x, \qquad (2.18)$$

where $\mathrm{E}^* \times \widetilde{\mathrm{E}}$ denotes the expectation with respect to the product measure $\mathrm{P}^* \times \widetilde{\mathrm{P}}$.

Let us note that the client's survival probability ${}_T p_x$ makes it impossible to hedge the payoff in (2.18) with probability one in view

$$0 < {}_T p_x < 1 \Rightarrow U_0 < H_0 = \mathrm{E}^* \left(H e^{-rT} \right). \qquad (2.19)$$

Relation (2.19) implies that the initial amount U_0 collected by the insurance company from the selling contract is strictly less that amount H_0, which is needed to hedge the payoff perfectly. In this case, quantile hedging (imperfect hedging technique), in general can be applied effectively to obtain an optimal hedging subject to a budget constraint.

Given an initial hedging capital U_0, we seek to find an admissible hedging strategy π^* that maximizes the probability of successful hedging

$$\mathrm{P}\left\{\omega : X_T^{\pi^*} \geq H\right\} = \max_\pi \mathrm{P}\left(\omega : X_T^\pi \geq H\right) \text{ with } V_0 \leq U_0 < H_0. \qquad (2.20)$$

The above inequalities reflect the fact that the investor is budget-constrained $U_0 < H_0$ and the requirement that the initial price V_0 of the optimal hedging strategy must not be greater than the amount available to the hedger $V_0 \leq U_0$. Therefore, the optimal hedging strategy π^* is the perfect hedge for the knock-out option $H \mathbb{1}_{A^*}$, where $A^* = \{1/Z_T > a^* e^{-rT} H\}$, $Z = Z_T$ is the density of the risk-neutral measure P^*, and $a^* = \inf\{a : \mathrm{Q}^* \{Z_T^{-1} > a e^{-T} H\} \leq a := U_0/H_0\}$.

Given that the amount U_0 is received from the policyholder, the insurance company will use quantile hedging results to find the perfect hedge π^* for the modified contingent claim $H \mathbb{1}_{A^*}$ with the initial price $V_0 = \mathrm{E}^* \left(H e^{-rT} \mathbb{1}_{A^*} \right) \leq U_0$. The strategy π^* will maximize the probability of successful hedging for H. These considerations allow us to obtain the following equation for the price of the equity-linked life insurance contract

$$U_0 = \mathrm{E}^* \left(H e^{-rT} \mathbb{1}_{A^*} \right) = \mathrm{E}^* \left(H e^{-rT} \right) {}_T p_x, \qquad (2.21)$$

and hence, we obtain a key balance equation

$${}_T p_x = \frac{\mathrm{E}^* \left(H \mathbb{1}_{A^*} \right)}{\mathrm{E}^* \left(H \right)}, \qquad (2.22)$$

where the right-hand side is called a hedging ratio.

Relations (2.21)-(2.22) are essential to the subsequent risk management analysis, as they give a quantitative connection between the financial and insurance risk components. Such connection, in turn, allows the insurance firm to assess accurately the risk it bears and implement specific strategies to control these risks according to the preferred risk management approach. That is, the firm can either offer an equity-linked life insurance contract for consideration to any client and then, based on the fair price received from this client, maximize the probability of successful hedging. Or, the firm can set the acceptable level of financial risk and then analyze clients for the contract accordingly.

More specifically, in the first approach, the client's survival probability $_Tp_x$ can be derived based on his/her known age x from some appropriate mortality model, allowing the firm to calculate $P(A^*)$. Note that the calculated maximal probability of successful hedging may not fit the company's desired risk profile. Alternatively, the firm can utilize (2.22) in reverse to choose some acceptable default risk level ϵ first, then find survival probabilities $_Tp_x$ for the potential clients. Next, using some particular mortality model, the ages of clients paying fair premiums (under the prescribed risk level) can be derived and risk management consequences can be analyzed in light of the firm's risk preferences.

We give our theoretical results in the following theorem.

Theorem 2.1. *Consider an insurance company that sells a single premium equity-liked life insurance contract with payoff $H_T = \max(S_T, K_T)$, the premium of the contract and the survival probability determined from quantile hedging is determined as follows. For the case $g > G$ and $\mu - r - \sigma^2 > 0$ the premium of the contract is*

$$U_0 = S_0 \Phi(\sigma\sqrt{T} - N^*) + S_0 \exp((g-r)T)\left[\Phi(N^*) - \Phi(M^*)\right] \quad (2.23)$$

and the survival probability is:

$$_Tp_x = 1 - \frac{\exp((g-r)T)\Phi(M^*)}{\Phi(\sigma\sqrt{T} - N^*) + \exp((g-r)T)\Phi(N^*)}. \quad (2.24)$$

For the case $g < G$ and $\mu - r - \sigma^2 < 0$, the premium of the contract is

$$\begin{aligned}U_0 = S_0 &\left[\Phi(\sigma\sqrt{T} - N^*) - \Phi(\sigma\sqrt{T} - J^*)\right] \\ &+ S_0 \exp((g-r)T)\left[\Phi(N^*) - \Phi(M^*)\right]\end{aligned} \quad (2.25)$$

and the survival probability is:

$$_Tp_x = 1 - \frac{\Phi(\sigma\sqrt{T} - J^*) + \exp((g-r)T)\Phi(M^*)}{\Phi(\sigma\sqrt{T} - N^*) + \exp((g-r)T)\Phi(N^*)}, \quad (2.26)$$

where $\Phi(j) = \int_{-\infty}^{j} \frac{\exp(\frac{-s^2}{2})}{\sqrt{2\pi}} ds$ is the cumulative distribution function for standard normal distribution, and $N^* = N + \theta\sqrt{T}$, $M^* = M + \theta\sqrt{T}$, $J^* =$

$$J + \theta\sqrt{T}, \quad \theta = \frac{\mu-r}{\sigma}, \quad G = \frac{\frac{\sigma^2}{T}\ln(S_0 a^*)}{\mu-r-\sigma^2} + \frac{\mu+r}{2}, \quad M = \frac{\left(g-r-\frac{\theta^2}{2}\right)T+\ln(S_0 a^*)}{\theta\sqrt{T}},$$

$$N = \frac{\left(g-\mu+\frac{\sigma^2}{2}\right)T}{\sigma\sqrt{T}}, \quad J = \frac{\ln(S_0 a^*) - \frac{(\theta-\sigma)^2}{2}T}{(\theta-\sigma)\sqrt{T}}.$$

Proof. Consider the financial model (2.12), (2.13) and the payoff of the contract (2.15). Using the general structure of the maximal set of successful hedging $A^* = \{1 > a^* Z_T \frac{H}{e^{rT}}\}$ we have

$$P(A^*) = E(\mathbb{1}_{A^*})$$
$$= E\left[\mathbb{1}_{\{1>Z_T \frac{S_T}{e^{rT}}\}}\mathbb{1}_{\{S_T \geq K_T\}}\right] + E\left[\mathbb{1}_{\{1>Z_T \frac{K_T}{e^{rT}}\}}\mathbb{1}_{\{S_T < K_T\}}\right]. \quad (2.27)$$

Since S and Z are both functions of W, and W_t is normally distributed with mean 0 and variance t (with respect to P), we can further rewrite (2.27) as

$$P(A^*) = E\left(\mathbb{1}_{\{y>J\}}\mathbb{1}_{\{y\geq N\}}\right) + E\left(\mathbb{1}_{\{y>M\}}\mathbb{1}_{\{y<N\}}\right) \text{ if } \mu-r-\sigma^2 > 0, \quad (2.28)$$

$$P(A^*) = E\left(\mathbb{1}_{\{y>J\}}\mathbb{1}_{\{y\geq N\}}\right) + E\left(\mathbb{1}_{\{y>M\}}\mathbb{1}_{\{y<N\}}\right) \text{ if } \mu-r-\sigma^2 < 0, \quad (2.29)$$

where y is a standard normal random variable (under P) and M, N, J are given in Theorem 2.1.

Next, we will compare M, N, J by performing some basic algebraic manipulations:

$$N - M = \frac{\frac{T}{\sigma}\left[(\mu-r)\left(g-\mu+\frac{\sigma^2}{2}\right) - \sigma^2\left(g-r+\frac{\sigma^2}{2}\right)\right] - \sigma\ln(a^* S_0)}{\sigma\theta\sqrt{T}} \quad (2.30)$$

$$= \frac{\Lambda}{\sigma\theta\sqrt{T}},$$

where $\Lambda = \frac{T}{\sigma}\left[(\mu-r)\left(g-\mu+\frac{\sigma^2}{2}\right) - \sigma^2\left(g-r+\frac{\sigma^2}{2}\right)\right] - \sigma\ln(a^* S_0)$.

Similarly,

$$N - J = \frac{\Lambda}{\sigma(\theta-\sigma)\sqrt{T}}, \quad M - J = \frac{\Lambda}{\theta(\theta-\sigma)\sqrt{T}}.$$

Now, observe that Λ is positive whenever

$$(\mu-r-\sigma^2)\left(g - \frac{\mu+r}{2}\right) > \frac{\sigma^2}{T}\ln(a^* S_0),$$

that is,

$$g > \frac{\frac{\sigma^2}{T}\ln(a^* S_0)}{\mu-r-\sigma^*} + \frac{\mu+r}{2}, \quad \mu-r-\sigma^2 > 0,$$

or

$$g < \frac{\frac{\sigma^2}{T}\ln(a^* S_0)}{\mu-r-\sigma^*} + \frac{\mu+r}{2}, \quad \mu-r-\sigma^2 < 0.$$

Moreover, based on the relationships between $N - M$, $N - J$, $M - J$, and the fact that the sign of $\theta - \sigma$ is determined by the sign of $\mu - r - \sigma^2$, if the guaranteed rate g is selected as indicated below, we obtain

$$J < M < N \text{ whenever } \mu - r - \sigma^2 > 0 \text{ and } g > G, \quad (2.31)$$

$$M < N < J \text{ whenever } \mu - r - \sigma^2 < 0 \text{ and } g < G, \quad (2.32)$$

where $G = \frac{\frac{\sigma^2}{T}\ln(a^*S_0)}{\mu - r - \sigma^*} + \frac{\mu + r}{2}$, which enables us to simplify P(A^*) even further. The two cases above produce the two different forms of A^* and the resulting expressions for the premium of the contracts (2.23) and (2.25).

Relations (2.31) and (2.32) lead to the following expressions for P(A^*)

$$P(A^*) = E\left(\mathbb{1}_{\{y > J\}}\mathbb{1}_{\{y \geq N\}}\right) + E\left(\mathbb{1}_{\{y > M\}}\mathbb{1}_{\{y < N\}}\right) = P\{y > M\}, \quad (2.33)$$

$$P(A^*) = E\left(\mathbb{1}_{\{y > J\}}\mathbb{1}_{\{y \geq N\}}\right) + E\left(\mathbb{1}_{\{y > M\}}\mathbb{1}_{\{y < N\}}\right) = P\{M < y < J\}, \quad (2.34)$$

from which we can obtain the representation for A^* : $A^* = \{y > M\}$ or $A^* = \{M < y < J\}$.

Using similar steps, we can simplify the expression for A^* under P* to

$$\begin{aligned} A^* &= \{y^* > J^*\} \cap \{y^* \geq N^*\} + \{y^* > M^*\} \cap \{y^* < N^*\} \\ &= \{y^* \geq N^*\} + \{M^* < y^* < N^*\} = \{y^* > M^*\}, \end{aligned} \quad (2.35)$$

or

$$\begin{aligned} A^* &= \{y^* > J^*\} \cap \{y^* \geq N^*\} + \{y^* > M^*\} \cap \{y^* < N^*\} \\ &= \{N^* \leq y^* < J^*\} + \{M^* < y^* < N^*\} = \{M^* < y^* < J^*\}, \end{aligned} \quad (2.36)$$

where the constants $N^* = N + \theta\sqrt{T}$, $M^* = M + \theta\sqrt{T}$, $J^* = J + \theta\sqrt{T}$.

Note that $J^* < M^* < N^*$ if $\mu - r - \sigma^2 > 0$ and $g > G$, or $M^* < N^* < J^*$ if $\mu - r - \sigma^2 < 0$ and $g < G$, and $N(0,1)$ under P*.

Recall that the expected logarithmic return of S under the risk-neutral measure P* is the risk-free interest rate r and hence

$$dS_t = S_t(rdt + \sigma dW_t^*), \quad (2.37)$$

where $W_t^* = W_t + \theta t$ is a Wiener process under P*.

We can rewrite U_0 as follows

$$\begin{aligned} U_0 &= E_*\left[\frac{H_T}{e^{rT}}\mathbb{1}_{A^*}\right] = E^*\left[\frac{S_T}{e^{rT}}\mathbb{1}_{\{1 > a^* Z_T \frac{S_T}{e^{rT}}\}}\mathbb{1}_{\{S_T \geq K_T\}}\right] \\ &\quad + E^*\left[\frac{K_T}{e^{rT}}\mathbb{1}_{\{1 > a^* Z_T \frac{K_T}{e^{rT}}\}}\mathbb{1}_{\{S_T < K_T\}}\right] \\ &= E^*\left[\frac{S_T}{e^{rT}}\mathbb{1}_{\{y^* > J^*\}}\mathbb{1}_{\{y^* \geq N^*\}}\right] + E^*\left[\frac{K_T}{e^{rT}}\mathbb{1}_{\{y^* > M^*\}}\mathbb{1}_{\{y^* < N^*\}}\right] \\ &= E^*\left[\frac{S_T}{e^{rT}}\mathbb{1}_{\{y^* \geq N^*\}}\right] + E^*\left[\frac{K_T}{e^{rT}}\mathbb{1}_{\{M^* < y^* < N^*\}}\right] \end{aligned} \quad (2.38)$$

for A^* in (2.35). Otherwise, for A^* in (2.36), we obtain a similar expression for U_0, which is omitted here but can be easily calculated as above.

Next, we proceed to calculate the explicit formula for the fair premium. The second term in the last line of (2.38) is simply

$$\mathrm{E}^* \left[\frac{K_T}{e^{rT}} \mathbb{1}_{\{M^* < y^* < N^*\}} \right] = S_0 \exp((g-r)T) \left[\Phi(N^*) - \Phi(M^*) \right].$$

The first term in the last line of (2.38) is calculated in a straightforward manner from (2.37) as

$$\mathrm{E}^* \left[\frac{K_T}{e^{rT}} \mathbb{1}_{\{y^* \geq N^*\}} \right] = S_0 \int_{N^*}^{+\infty} e^{-\frac{1}{2}\sigma^2 T} e^{\sigma\sqrt{T} y^*} e^{-\frac{1}{2}(y^*)^2} \frac{1}{\sqrt{2\pi}} dy^*$$

$$= S_0 \Phi(\sigma\sqrt{T} - N^*).$$

□

Let us discuss the conditions of the above theorem.

An insurance company offering the contract can choose an appropriate guaranteed rate g based on the market situation and the nature of the risky asset expressed by the relations between parameters μ, r and σ. We should always have $r < g < \mu$ because the guaranteed rate should be higher than the risk-free rate, otherwise clients would find money markets more appealing. Further, g should also be lower than the expected return on the risky asset, since payment of the guarantee involves no risk. The condition $\mu - r - \sigma^2 > 0$ implies that the expected excess return of the risky asset over the risk-free rate is higher than the risk (as volatility σ^2) associated with the risky asset. Then the stock is more attractive than the guarantee, whose parameters satisfy $\mu - r - \sigma^2 < 0$, since the risk-return relation for the guarantee is not as appealing as in the previous case. The only restriction is that $\mu - r - \sigma^2 \neq 0$, since this term appears in the denominator during calculations. However, generally speaking, this term should not pose any problems, since finding risky assets satisfying the precise relationship $\mu - r = \sigma^2$ would likely prove difficult.

A simple guideline for a manager deciding how to set guaranteed rates for constants involving stock and guarantee is the following. To guarantee the higher expected return of the more appealing stock, the manager should set g to exceed the average of the asset's return and the risk-free rate. That is, for $\mu - r - \sigma^2 > 0$, $\frac{\mu+r}{2} < g < \mu$. Note, that this ensures $g > G$, as required by conditions in relation (2.31). If on the other hand, the underlying equity of the contract is less attractive than the guarantee, the manager can set the guarantee below the average of μ and r ($r < g < \frac{\mu+r}{2}$ implies $g < G$ for relation (2.32)). That is, whenever the contract calls for securing an asset whose expected return is not very high, the guaranteed rate can be lower than in the case of an underlying equity with high returns.

2.2 Quantile hedging of contracts with stochastic guarantees

In our model, the two assets involved are based on different Wiener processes, thus the movements in the asset price are not perfectly correlated. Such a generalized model creates a much more realistic setting for future actuarial analysis. We have a financial market with a nonzero interest rate and two risky assets, of which the first one, S^1, determines the size of potential future gains to the holder of the contract, while the more reliable second asset, S^2, serves as a flexible guarantee for the insured. We describe the setting in detail, discuss why quantile hedging applies to our case, and find the price of the contract under consideration using quantile hedging and the Margrabe formula.

We have a financial market with interest rate $r \geqslant 0$, so the evolution of the non-risky asset B_t (such as money in the bank account) is given by

$$dB_t = rB_t dt, \quad B_0 \equiv 1.$$

We have two risky assets, S^1 and S^2, which are based on different (correlated) Wiener processes W_t^1 and W_t^2. The evolution of asset prices is given by the following stochastic differential equations:

$$dS_t^i = S_t^i(\mu_i dt + \sigma_t dW_t^i), \quad i = 1, 2, \qquad (2.39)$$

with $\mu_i \in \mathbb{R}$, $\sigma_i > 0$.

We assume that the second asset S^2 is "safer" that the first one, so that $\sigma_1 > \sigma_2$. The covariance of W_t^1 and W_t^2 is $\rho \cdot t$, where $\rho^2 < 1$.

All processes are given on a standard stochastic basis $(\Omega, \mathcal{F}, \mathbb{F} = (\mathcal{F}_t)_{t \geqslant 0}, P)$ and are adapted to the filtration $\mathbb{F}$, generated by W_t^i.

Every predictable process $\pi = (\pi_t)_{t \geqslant 0} = (\beta_t, \gamma_t^1, \gamma_t^2)_{t \geqslant 0}$ is called a trading strategy, or a portfolio. The capital (discounted capital) of π at time t equals

$$X_t^\pi = \beta_t B_t + \gamma_t^1 S_t^1 + \gamma_t^2 S_t^2 \Leftrightarrow$$
$$\frac{X_t^\pi}{B_t} = \beta_t + \gamma_t^1 \frac{S_t^1}{B_t} + \gamma_t^2 \frac{S_t^2}{B_t}. \qquad (2.40)$$

The strategies whose capital satisfies

$$\frac{X_t^\pi}{B_t} = X_0^\pi + \sum_{i=1}^{2} \int_0^t \gamma_u^i d\left(\frac{S_u^i}{B_u}\right) \qquad (2.41)$$

are called self-financing, and the class of such strategies is denoted SF. Only $\pi \in SF$ with nonnegative capital are considered admissible.

It is well known that in order to prevent arbitrage, we must work with the probability measure that allows no riskless profits. Traditionally, this risk-neutral measure is denoted P*. Under P*, the returns of assets S^1 and S^2 are equal to the risk-free interest rate r, and prices calculated w.r. to P* are arbitrage-free.

We calculate the density $Z_t = \frac{dP_t^*}{dP_t}$ of the martingale measure P*, using the general methodology briefly described in Chapter 1. Let us express Z as a stochastic exponent of some martingale process N:

$$Z_t = \varepsilon(N_t).$$

Since we have two Brownian motions, we should write N_t in the form $N_t = \phi_1 \cdot W_t^1 + \phi_2 \cdot W_t^2$.

We can represent B_t, S_t^1 and S_t^2 as stochastic exponents of processes h, H_t^1 and H_t^2, respectively:

$$h = rt, \quad H_t^i = \mu_i t + \sigma_i W_t^i.$$

The general methodology for finding martingale measures states that the process

$$\Psi(h, H, N) = H_t^i - h_t + N_t + \left\langle (h - H^i)^c, (h - N)^c \right\rangle_t$$

should be a martingale w.r. to P, and from this we can find constants ϕ_1 and ϕ_2.

For Ψ_1 and Ψ_2, we get the following expressions:

$$\Psi_1 = \mu_1 t + \sigma_1 W_t^1 - rt + \phi_1 W_t^1 + \phi_2 W_t^2 + \sigma_1 \phi_1 t + \sigma_1 \phi_2 \rho t,$$

$$\Psi_2 = \mu_2 t + \sigma_2 W_t^2 - rt + \phi_1 W_t^1 + \phi_2 W_t^2 + \sigma_2 \phi_2 t + \sigma_2 \phi_1 \rho t.$$

To make these processes by martingales, we must have

$$\mu_1 t - rt + \sigma_1 \phi_1 t + \sigma_1 \phi_2 \rho t = 0 \text{ and } \mu_2 t - rt + \sigma_2 \phi_2 t + \sigma_2 \phi_1 \rho t = 0,$$

and therefore

$$\phi_1 = \frac{r(\sigma_2 - \sigma_1 \rho) + \rho \mu_2 \sigma_1 - \mu_1 \sigma_2}{\sigma_1 \sigma_2 (1 - \rho^2)}$$

and (2.42)

$$\phi_2 = \frac{r(\sigma_1 - \sigma_2 \rho) + \rho \mu_1 \sigma_2 - \mu_2 \sigma_1}{\sigma_1 \sigma_2 (1 - \rho^2)}.$$

Now, going back to the stochastic exponent form, we get the following expression for Z:

$$Z_t = \frac{dP_t^*}{dP_t} = \varepsilon(N_t) = \varepsilon(\phi_1 W_t^1 + \phi_2 W_t^2)$$

$$= \exp\left\{\phi_1 W_t^1 + \phi_2 W_t^2 - \frac{1}{2}(\phi_1^2 + \phi_2^2 + 2\rho \phi_1 \phi_2) t\right\}.$$

(2.43)

Note that in our setup the martingale measure P* is unique, and the market (2.39) is complete.

Now, as we mentioned before, S_t^1 and S_t^2 are martingales w.r. to P*, which we now show. Consider the following two processes:

$$W_t^{1*} = W_t^1 + t \cdot \frac{\mu_1 - r}{\sigma_1}$$

and (2.44)

$$W_t^{2*} = W_t^2 + t \cdot \frac{\mu_2 - r}{\sigma_2}.$$

One can check that W^{i*} are new Wiener processes under P*, and that their covariance is $\rho \cdot t$.

With the help of (2.44), we rewrite (2.39) and represent S^1 and S^2 as

$$dS_t^1 = S_t^1(\mu_1 dt + \sigma_1 dW_t^1) = S_t^1(rdt + \sigma_1 dW_t^{1*}) \Leftrightarrow$$

$$S_t^1 = S_0^1 \cdot \exp\left\{rt - \frac{\sigma_1^2}{2}t + \sigma_1 W_t^{1*}\right\}$$

and (2.45)

$$dS_t^2 = S_t^2(\mu_2 dt + \sigma_2 dW_t^2) = S_t^2(rdt + \sigma_2 dW_t^{2*}) \Leftrightarrow$$

$$S_t^2 = S_0^2 \cdot \exp\left\{rt - \frac{\sigma_2^2}{2}t + \sigma_2 W_t^{2*}\right\}.$$

Let us fix a time horizon T. Any nonnegative $\mathcal{F}_T$-measurable random variable H will be called a contingent claim. Let us take a strategy π and form its value starting from an initial capital $x = X_0^\pi$. We call $A(x, \pi)$ the set of successful hedging if

$$A(x, \pi) = \{\omega : X_T^\pi \geqslant H\}.$$

It follows from option pricing theory that, in a complete market, there exists a strategy π^* with the property

$$P(A(\mathrm{E}^* e^{-rT} H, \pi^*)) = 1, \qquad (2.46)$$

where $X_0^\pi = X_0 = \mathrm{E}^* e^{-rT} H$, and π^* is a perfect hedge.

However, often $X_0 < \mathrm{E}^* e^{-rT} H$, and cannot provide the financing needed for the perfect hedge in the sense of (2.46). Then how do we choose the optimal hedging method for the given contingent claim with limited initial capital? One natural answer is to maximize the probability of successful hedging: choose a strategy π^* such that

$$P(A(x, \pi^*)) = \max_\pi P(A(x, \pi)), \qquad (2.47)$$

under the restriction $x \leqslant X_0 < \mathrm{E}^* H e^{-rT}$.

As we know from Paragraph 2.1, the solution set to (2.47) $A^* = A(X_0, \pi^*)$, called the maximal set of successful hedging, has the following structure:

$$A^* = \{Z_T^{-1} \geq a \cdot H\}, \tag{2.48}$$

where optimal strategy π^* is a perfect hedge for the modified claim

$$H_{A^*} = H \mathbb{1}_{A^*}. \tag{2.49}$$

The maximization problem in (2.47) and the structure of A^* in (2.48) have a statistical flavor connected with the fundamental Neyman–Pearson lemma. The type of hedging methodology described in Paragraph 2.1 and here is called quantile hedging.

We just mentioned that sometimes the writer of the contract does not receive sufficient funds to hedge the contract perfectly: $x \leq X_0 < \mathrm{E}^* e^{-rT} H$. In our case, the factor that causes this to happen is the mortality of the client (the insurance component of risk).

Following actuarial traditions, we use a random variable $T(x)$ on some probability space $(\widetilde{\Omega}, \widetilde{\mathcal{F}}, \widetilde{\mathrm{P}})$ to represent the remaining lifetime of an insured of age x. Let us consider a pure endowment contract with the payoff function

$$H(T(x)) = H \cdot \mathbb{1}_{\{T(x) > T\}}. \tag{2.50}$$

To find a natural value of X_0 called the Brennan–Schwartz price we take the expected value in (2.50) w. r. to $\mathrm{P}^* \times \widetilde{\mathrm{P}}$:

$$\mathrm{E}^* \times \widetilde{\mathrm{E}}(e^{-rT} H(T(x))) = \mathrm{E}^* e^{-rT} H \widetilde{\mathrm{E}} \mathbb{1}_{\{T(x) > T\}} = \mathrm{E}^* H\, _T p_x, \tag{2.51}$$

where $_T p_x = \widetilde{\mathrm{P}}(T(x) > T)$.

In view of (2.49), the appropriate value of the initial capital of the hedging portfolio for this contingent claim should be

$$X_0 = \mathrm{E}^* H \cdot {_T p_x} < \mathrm{E}^* H e^{-rT}. \tag{2.52}$$

Obviously, the price that the client should pay for the contract with the payoff (2.50), (the initial capital needed for hedging) is less than the amount the insurance company should invest to guarantee a perfect hedge. The condition (2.52) shows us that in order to hedge its potential liability with maximal probability, the company should apply quantile hedging methodology (2.47)–(2.49). We show how to do this for the contract with the payoff $\max\{S_t^1, S_T^2\} \mathbb{1}_{\{T(x) > T\}}$, and along with quantile hedging, we will use a very helpful Margrabe formula for the price of the option with the payoff $(S_T^1 - S_T^2)^+$:

$$\mathrm{E}^*(S_T^1 - S_T^2)^+ = \mathcal{C}^{\mathrm{Mar}}(S^1, S^2, T) = S_0^1 \Phi(b_+) - S_0^2 \Phi(b_-), \tag{2.53}$$

$$b_\pm = \frac{\ln \frac{S_0^1}{S_0^2} \pm \frac{\sigma^2}{2} T}{\sigma \sqrt{T}},$$

$$\sigma^2 = \sigma_1^2 + \sigma_2^2 - 2\sigma_1\sigma_2\rho,$$

$$\Phi(x) = \frac{1}{\sqrt{2\pi}} \int_{-\infty}^{x} \exp\left\{-\frac{y^2}{2}\right\} dy.$$

Now, consider a life insurance contract, whose payoff $\max\{S_T^1, S_T^2\}$ is conditional upon the client's survival to maturity of the contract, T. The actual payoff for this contract is

$$\max\{S_T^1, S_T^2\} \cdot \mathbb{1}_{\{T(x)>T\}}. \tag{2.54}$$

Here, $T(x)$ is the remaining life of the insured of age x, as defined in (2.50). The first riskier asset S^1 is responsible for potentially large gains to the holder of the contract, while the safer S^2 guarantees that the insured will receive at least the amount S_t^2 (that is, if he/she lives to collect the money).

Note that

$$\max\{S_T^1, S_T^2\} = S_T^2 + (S_T^1 - S_T^2)^+,$$

and under the martingale measure P* we obtain

$$\begin{aligned}\mathrm{E}^*(e^{-rT} \max\{S_T^1, S_T^2\}) &= \mathrm{E}^*(e^{-rT} S_T^2) + \mathrm{E}^*(e^{-rT}(S_T^1 - S_T^2)^+) \\ &= S_0^2 + \mathrm{E}^*(e^{-rT}(S_T^1 - S_T^2)^+).\end{aligned}$$

We know that the initial price of a contract is the expectation w.r. to the risk-neutral measure (P* in our case) of the contract's discounted payoff. Thus, taking into account the above equation and (2.51), we calculate the appropriate initial price $_T U_x$ of our contract (2.54) as follows

$$\begin{aligned}_T U_x &= \mathrm{E}^* \times \widetilde{\mathrm{E}}\left(e^{-rT} \max\{S_T^1, S_T^2\} \mathbb{1}_{\{T(x)>T\}}\right) \\ &= \mathrm{E}^*\left(e^{-rT} \max\{S_T^1, S_T^2\}{}_T p_x\right) \\ &= S_0^2 {}_T p_x + \mathrm{E}^*\left(e^{-rT}(S_T^1 - S_T^2)^+\right){}_T p_x.\end{aligned} \tag{2.55}$$

The above equation shows that in order to find the initial price of (2.54) it is sufficient to calculate the price of $e^{-rT}(S_T^1 - S_T^2)^+ {}_T p_x$. Therefore we will work with this more convenient payoff, called the embedded options.

Now, $(S_T^1 - S_T^2)^+ {}_T p_x < (S_T^1 - S_T^2)^+$ implies that

$$X_0 = e^{-rT} \mathrm{E}^*(S_T^1 - S_T^2)^+ {}_T p_x < e^{-rT} \mathrm{E}^*(S_T^1 - S_T^2)^+,$$

that is, the initial value of our contract (the price that an insurance company would receive) is less than the amount required for the perfect hedge. Thus, as we conclude in (2.50)–(2.52), quantile hedging is an appropriate method to price the above contract.

From the option pricing theory, we know that the price is the expectation of the payoff w.r. to the risk-neutral measure P*, while quantile hedging theory tells us to calculate this price as the initial value of the perfect hedge for the modified payoff (see (2.49)). Therefore, we get this equality for the price of the contract

$$\mathrm{E}^*(e^{-rT}(S_T^1 - S_T^2)^+)_T p_x = \mathrm{E}^*\left(e^{-rT}(S_T^1 - S_T^2)^+ \mathbf{1}_{A^*}\right), \qquad (2.56)$$

which, in turn, leads to the following expression for the survival probability $_T p_x$:

$$_T p_x = \frac{\mathrm{E}^*\left((S_T^1 - S_T^2)^+ \mathbf{1}_{A^*}\right)}{\mathrm{E}^*\left(S_T^1 - S_T^2\right)^+}, \qquad (2.57)$$

where A^* is the maximal set of successful hedging for $(S_T^1 - S_T^2)^+$.

The balance equation (2.57) is essential to our future actuarial analysis, because it quantifies the relationship between the insurance and the financial components of the contract's risk. Moreover, (2.57) shows that if an insurance company takes on too much financial risk, it can compensate by allowing less insurance risk (signing contracts with older clients), and vice versa.

Now, let us move on to calculating the initial price. To do this, we rewrite the representation for A^* in (2.48) as follows:

$$\begin{aligned} A^* &= \left\{ Z_T^{-1} \geq a \left(S_T^1 - S_T^2\right)^+ \right\} = \left\{ Z_T^{-1} \geq a S_T^2 \left(\frac{S_T^1}{S_T^2} - 1\right)^+ \right\} \\ &= \left\{ (Z_T S_T^2)^{-1} \geq a \left(\frac{S_T^1}{S_T^2} - 1\right)^+ \right\}, \end{aligned} \qquad (2.58)$$

where a is some appropriate constant.

The representation (2.58) shows that we should work with the ratio $Y_T = \frac{S_T^1}{S_T^2}$.

Taking into account (2.45), we have for Y_T:

$$Y_T = \frac{S_T^1}{S_T^2} = \frac{S_0^1}{S_0^2} \exp\left\{-\frac{(\sigma_1^2 - \sigma_2^2)}{2}T + \sigma_1 W_T^{1*} - \sigma_2 W_T^{2*}\right\}. \qquad (2.59)$$

To simplify further calculations, we note here that

$$W_t^* = \frac{\sigma_1 W_t^{1*} - \sigma_2 W_t^{2*}}{\sigma} \qquad (2.60)$$

is a new Brownian motion w.r. to P*, with σ^2 defined in (2.53). One can calculate the corresponding covariances to equal

$$\begin{aligned} \mathrm{cov}(W_t^*, W_t^{1*}) &= \mathrm{E}^*(W_t^* \cdot W_t^{1*}) = \frac{(\sigma_1 - \sigma_2 \rho)t}{\sigma}, \\ \mathrm{cov}(W_t^*, W_t^{2*}) &= \mathrm{E}^*(W_t^* \cdot W_t^{2*}) = \frac{(\sigma_1 \rho - \sigma_2)t}{\sigma}. \end{aligned} \qquad (2.61)$$

Using (2.60), we express Y_T in (2.59) as a function of a single Brownian motion:

$$Y_T = \frac{S_0^1}{S_0^2} \exp\left\{-\frac{(\sigma_1^2 - \sigma_2^2)}{2}T + \sigma W_T^*\right\}. \qquad (2.62)$$

We will use of the form (2.62) of Y_T eventually.

Now let us go back to the maximal set of successful hedging A^*. We will try to express A^* in terms of Y_T and some constant functions. First, we find $Z_T \cdot S_T^2$ as a function of Y_T (see (2.58)).

Using (2.44), we rewrite the expression for Z as follows:

$$Z_T = \exp\left\{\phi_1 W_T^1 + \phi_2 W_T^2 - \frac{1}{2}\left(\phi_1^2 + \phi_2^2 + 2\rho\phi_1\phi_2\right)T\right\}$$

$$= \exp\left\{\phi_1\left(W_T^{1*} - T \cdot \frac{\mu_1 - r}{\sigma_1}+\right) + \phi_2\left(W_T^{2*} - T \cdot \frac{\mu_2 - r}{\sigma_2}+\right)\right\}$$

$$\times \exp\left\{-\frac{1}{2}(\phi_1^2 + \phi_2^2 + 2\rho\phi_1\phi_2)T\right\} \qquad (2.63)$$

$$= \exp\left\{\phi_1 W_T^{1*} + \phi_2 W_T^{2*}\right\}$$

$$\times \exp\left\{\left(-\frac{\phi_1(\mu_1 - r)}{\sigma_1} + \frac{\phi_2(\mu_2 - r)}{\sigma_2} + \frac{1}{2}(\phi_1^2 + \phi_2^2 + 2\rho\phi_1\phi_2)\right)T\right\}.$$

Then, (2.45) and (2.60) give us

$$Z_T S_T^2 = \exp\left\{\phi_1 W_T^{1*} - T \cdot \frac{\phi_1(\mu_1 - r)}{\sigma_1} + \phi_2 W_T^{2*} - T \cdot \frac{\phi_2(\mu_2 - r)}{\sigma_2}\right\}$$

$$\times \exp\left\{-\frac{1}{2}(\phi_1^2 + \phi_2^2 + 2\rho\phi_1\phi_2)T\right\}$$

$$\times S_0^2 \exp\left\{rT - \frac{\sigma_2^2}{2}T + \sigma_2 W_t^{2*}\right\}$$

$$= S_0^2 \exp\left\{\phi_1 W_T^{1*} + (\phi_2 + \sigma_2)W_t^{2*}\right\}$$

$$\times \exp\left\{rT - \frac{\sigma_2^2}{2}T - \frac{\phi_1(\mu_1 - r)}{\sigma_1}T - \frac{\phi_2(\mu_2 - r)}{\sigma_2}T\right\}$$

$$\times \exp\left\{-\frac{1}{2}(\phi_1^2 + \phi_2^2 + 2\rho\phi_1\phi_2)T\right\} \qquad (2.64)$$

$$= \left(\frac{S_0^1}{S_0^2}\right)^\alpha \exp\left\{\sigma_1 \alpha W_T^{1*} - \sigma_2 \alpha W_T^2 - T\alpha\frac{\sigma_1^2 - \sigma_2^2}{2}\right\}$$

$$\times \left(\frac{S_0^2}{S_0^1}\right)^\alpha \exp\left\{T\alpha\frac{\sigma_1^2 - \sigma_2^2}{2}\right\}$$

$$\times S_0^2 \exp\left\{rT - \frac{\sigma_2^2}{2}T - \frac{\phi_1(\mu_1 - r)}{\sigma_1} - \frac{\phi_2(\mu_2 - r)}{\sigma_2}\right\}$$

$$\times \exp\left\{-\frac{1}{2}(\phi_1^2 + \phi_2^2 + 2\rho\phi_1\phi_2)T\right\}$$

$$= Y_T^\alpha \cdot g,$$

where α is our work parameter different from the parameter of Lemma 2.1,

$$g = \frac{(S_0^2)^{\alpha+1}}{(S_0^1)^{\alpha}} \exp\left\{T\alpha\left(\frac{\sigma_1^2 - \sigma_2^2}{2}\right) + rT - \frac{\sigma_2^2}{2}T\right\}$$

$$\times \exp\left\{-\left(\frac{\phi_1(\mu_1 - r)}{\sigma_1} + \frac{\phi_2(\mu_2 - r)}{\sigma_2} + \frac{1}{2}(\phi_1^2 + \phi_2^2 + 2\rho\phi_1\phi_2)\right)T\right\}.$$
(2.65)

From (2.64), we must have

$$\sigma_1 \cdot \alpha = \phi_1 \quad \text{and} \quad -\sigma_2 \cdot \alpha = \phi_2 + \sigma_2, \qquad (2.66)$$

which leads to the following condition for our model:

$$\alpha = \frac{\phi_1}{\sigma_1} = -1 - \frac{\phi_2}{\sigma_2}. \qquad (2.67)$$

Now we have achieved our goal of representing A^* in terms of Y: from (2.58) and (2.64)

$$A^* = \left\{\frac{1}{Y_T^{\alpha}} \geq g \cdot \alpha(Y_T - 1)^+\right\}. \qquad (2.68)$$

To analyze A^* in the form (2.68) consider the following characteristic equation

$$x^{-\alpha} = g \cdot \alpha \cdot (x - 1)^+. \qquad (2.69)$$

Depending on the value of α, the function $x^{-\alpha}$ will behave differently, so it is useful to look at three possibilities separately.

Case 1: $-\alpha \leq 0$;
Case 2: $0 < -\alpha \leq 1$;
Case 3: $-\alpha > 1$.

In the first two cases, (2.69) has one solution $c(\alpha)$, while in the third case there are two solutions $c_1(\alpha) < c_2(\alpha)$ (or none at all). From this, we conclude that A^* is equivalent to $\{Y_T \leq c(\alpha)\}$ for cases 1 and 2, and for cases 3 it becomes $\{Y_T \leq c_1(\alpha)\} \bigcup \{Y_T \geq c_2(\alpha)\}$.

To find the final formula for the price $\mathcal{C}$ of our contingent claim (see (2.56)), we consider cases 1 and 2 first.

Keeping in mind that because $c(\alpha) \geq 1$,

$$(S_T^1 - S_T^2)^+ \mathbb{1}_{\{Y_T \leq c(\alpha)\}} = (S_T^1 - S_T^2)^+ - S_T^1 + S_T^2 + (S_T^1 - S_T^2)\mathbb{1}_{\{Y_T \leq c(\alpha)\}},$$

we can write the following equivalences for the price of the contract:

$$\mathcal{C} = \mathbf{E}^* \frac{(S_T^1 - S_T^2)^+ {}_T p_x}{e^{rT}} = \mathbf{E}^* \frac{(S_T^1 - S_T^2)^+}{e^{rT}} \mathbb{1}_{A^*}$$

$$= \mathbf{E}^* \frac{(S_T^1 - S_T^2)^+}{e^{rT}} \mathbb{1}_{\{Y_T \leq c(\alpha)\}} \qquad (2.70)$$

$$= \mathbf{E}^* \frac{(S_T^1 - S_T^2)^+}{e^{rT}} - \mathbf{E}^* \frac{S_t^1}{e^{rT}} + \mathbf{E}^* \frac{S_t^2}{e^{rT}} + \mathbf{E}^* \frac{(S_T^1 - S_T^2)}{e^{rT}} \mathbb{1}_{\{Y_T \leq c(\alpha)\}}.$$

Let us look at (2.70) in the last line above. We know how to calculate the first term, $E^*(S_T^1 - S_T^2)^+$; its value is given by Margrabe's formula (2.53). Since both stocks are martingales w.r. to P^*, $E^* \frac{S_T^1}{e^{rT}} = S_0^1$ and $E^* \frac{S_T^2}{e^{rT}} = S_0^2$. So the only unknown is the last term

$$E^* \frac{(S_T^1 - S_T^2)}{e^{rT}} \mathbb{1}_{\{Y_T \leq c(\alpha)\}}. \tag{2.71}$$

To find it, we will apply the following probabilistic lemma.

Lemma 2.3. *For $i = 1, 2$ and two normally distributed random variables $\eta_i \sim N(\mu_{\eta_i}, \sigma_{\eta_i}^2)$ and $\zeta \sim N(\mu_\zeta, \sigma_\zeta^2)$,*

$$E^* \left(\exp\{-\eta_i\} \mathbb{1}_{\{\zeta \leq \ln(c)\}} \right)$$
$$= \exp\left\{ \frac{\sigma_{\eta_i}^2}{2} - \mu_{\eta_i} \right\} \Phi\left(\frac{\ln(c) - (\mu_\zeta - \mathrm{cov}(\zeta, \eta_i))}{\sigma_\zeta} \right). \tag{2.72}$$

We rewrite the set in (2.71) as follows

$$\{Y_T \leq c(\alpha)\} = \{\ln(Y_T) \leq \ln(c(\alpha))\},$$

and denote

$$\zeta = \ln(Y_T), \quad S_T^i = \exp\{-\eta_i\},$$

where the Gaussian random variables

$$\eta_i = -\ln(S_0^i) - rT + \frac{\sigma_i^2}{2} T - \sigma_i W_T^{i*}, \quad i = 1, 2, \tag{2.73}$$

are defined in (2.44)–(2.45).

The next step is finding the parameters used in Lemma 2.3,

$$\mu_\zeta = \mu_{\ln(Y_T)} = E^*(\ln(Y_T))$$
$$= E^* \left(\ln\left(\frac{S_0^1}{S_0^2} \exp\left\{ -\frac{\sigma_1^2 - \sigma_2^2}{2} T + \sigma W_T^* \right\} \right) \right)$$
$$= E^* \left(\ln\left(\frac{S_0^1}{S_0^2} \right) - \frac{\sigma_1^2 - \sigma_2^2}{2} T + \sigma W_T^* \right) = \ln\left(\frac{S_0^1}{S_0^2} \right) - \frac{\sigma_1^2 - \sigma_2^2}{2} T,$$

$$\sigma_\zeta^2 = E^*(\zeta - \mu_\zeta)^2$$
$$= E^* \left(\ln\left(\frac{S_0^1}{S_0^2} \right) - \frac{\sigma_1^2 - \sigma_2^2}{2} T + \sigma W_T^* - \left[\ln\left(\frac{S_0^1}{S_0^2} \right) - \frac{\sigma_1^2 - \sigma_2^2}{2} T \right] \right)^2$$
$$= \sigma^2 T,$$

$$\mu_{\eta_i} = E^* \left(-\ln(S_0^i) - rT + \frac{\sigma_i^2}{2} T - \sigma_i W_T^{i*} \right) = -\ln(S_0^i) - rT + \frac{\sigma_i^2}{2} T,$$

$$\sigma_{\eta_i}^2 = \mathrm{E}^*(\eta_i - \mu_{\eta_i})^2$$
$$= \mathrm{E}^*\left(-\ln(S_0^i) - rT + \frac{\sigma_i^2}{2}T - \sigma_i W_T^{i*} - \left[-\ln(S_0^i) - rT + \frac{\sigma_i^2}{2}T\right]\right)$$
$$= \sigma_i^2 T,$$

$$\mathrm{cov}(\zeta, \eta_i) = \mathrm{E}^*((\zeta - \mu_\zeta)(\eta_i - \mu_{\eta_i})) = \mathrm{E}^*(\sigma W_T^*(-\sigma_i W_T^{i*}))$$
$$= -\sigma\sigma_i \mathrm{cov}(W_T^*, W_T^{i*}), \quad i = 1, 2,$$

with $\mathrm{cov}(W_T^*, W_T^{i*})$ given by (2.61).

Substituting the above values into (2.72), we calculate (2.71):

$$\mathrm{E}^*\left(\frac{S_T^1 - S_T^2}{e^{rT}} \mathbb{1}_{\{Y_T \leq c(\alpha)\}}\right) = \mathrm{E}^*\left(\frac{S_T^1}{e^{rT}} \mathbb{1}_{\{Y_T \leq c(\alpha)\}}\right) - \mathrm{E}^*\left(\frac{S_T^2}{e^{rT}} \mathbb{1}_{\{Y_T \leq c(\alpha)\}}\right);$$

$$\mathrm{E}^*\left(\frac{S_T^1}{e^{rT}} \mathbb{1}_{\{\ln(Y_T) \leq \ln(c(\alpha))\}}\right) = e^{-rT} e^{\frac{\sigma_1^2}{2}T - \frac{\sigma_1^2}{2}T + \ln(S_0^1) + rT}$$
$$\times \Phi\left(\frac{\ln(c(\alpha)) - \left(\ln\left(\frac{S_0^1}{S_0^2}\right) - \frac{\sigma_1^2 - \sigma_2^2}{2}T + \sigma_1(\sigma_1 - \sigma_2\rho)T\right)}{\sigma\sqrt{T}}\right)$$
$$= S_0^1 \Phi\left(\frac{-\left[\ln\frac{S_0^1}{c(\alpha)S_0^2} + \frac{\sigma^2}{2}T\right]}{\sigma\sqrt{T}}\right);$$

$$\mathrm{E}^*\left(\frac{S_T^2}{e^{rT}} \mathbb{1}_{\{\ln(Y_T) \leq \ln(c(\alpha))\}}\right) = e^{\frac{\sigma_2^2}{2}T - \frac{\sigma_2^2}{2}T + \ln(S_0^2)}$$
$$\times \Phi\left(\frac{\ln(c(\alpha)) - \left(\ln\left(\frac{S_0^1}{S_0^2}\right) - \frac{\sigma_1^2 - \sigma_2^2}{2}T + \sigma_2(\sigma_1\rho - \sigma_2)T\right)}{\sigma\sqrt{T}}\right)$$
$$= S_0^2 \Phi\left(\frac{-\left[\ln\frac{S_0^1}{c(\alpha)S_0^2} - \frac{\sigma^2}{2}T\right]}{\sigma\sqrt{T}}\right).$$

Therefore, the initial price of the payoff in (2.56) equals

$$\mathcal{C} = \mathcal{C}^{\mathrm{Mar}} - S_0^1 + S_0^2 + S_0^1 \Phi(-b_+(c)) - S_0^2 \Phi(-b_-(c))$$
$$= S_0^1 \Phi(b_+(1)) - S_0^2 \Phi(b_-(1)) - \left(S_0^1 \Phi(b_+(c)) - S_0^2 \Phi(b_-(c))\right)$$
$$= S_0^1 \left[\Phi(b_+(1)) - \Phi(b_+(c))\right] - S_0^2 \left[\Phi(b_-(1)) - \Phi(b_-(c))\right];$$

$$b_+(c) = \frac{\ln \frac{S_0^1}{c(\alpha)S_0^2} + \frac{\sigma^2}{2}T}{\sigma\sqrt{T}},$$
$$b_-(c) = \frac{\ln \frac{S_0^1}{c(\alpha)S_0^2} - \frac{\sigma^2}{2}T}{\sigma\sqrt{T}},$$
(2.74)

with σ given in (2.53). To simplify this formula, we use the fact that b_+ and b_- in Margrabe's formula (2.53) are actually $b_+(1)$ and $b_-(1)$, according to the above definition of $b_\pm(c)$.

It is worthwhile to point out that the explicit price formula (2.74) is independent of the interest rate r, as is Margrabe's result, even thought r shows up in the expression for the density Z and the representation of A^* in the constant g (2.65).

Having calculated the initial price, we can derive the formula for the survival probability (2.57) for cases 1 and 2:

$$_T p_x = 1 - \frac{S_0^1 \Phi(b_+(c)) - S_0^2 \Phi(b_-(c))}{S_0^1 \Phi(b_+(1)) - S_0^2 \Phi(b_-(1))}, \qquad (2.75)$$

with $b_\pm(c)$ given by (2.74).

Then for the price of the original contract, (2.54) becomes (2.55)

$$\begin{aligned}_T U_x &= \mathrm{E}^*(e^{-rT}(S_T^1 - S_T^2)^+)_T p_x + S_0^2{}_T p_x \\ &= \left[1 - \frac{S_0^1 \Phi(b_+(c)) - S_0^2 \Phi(b_-(c))}{S_0^1 \Phi(b_+(1)) - S_0^2 \Phi(b_-(1))}\right] \\ &\quad \times \left[S_0^1 \Phi(b_+(1)) - S_0^2 \Phi(b_-(1)) + S_0^2\right].\end{aligned} \qquad (2.76)$$

Let us return to case 3, with $-\alpha > 1$, and approach the price calculations in the following way.

Note that the solution sets $\{Y_T \leqslant c_1(\alpha)\}$ and $\{Y_T \geqslant c_2(\alpha)\}$ are disjoint, thus

$$\mathbb{1}_{\{Y_T \leqslant c_1(\alpha)\} \cup \{Y_T \geqslant c_2(\alpha)\}} = \mathbb{1}_{\{Y_T \leqslant c_1(\alpha)\}} + \mathbb{1}_{\{Y_T \geqslant c_2(\alpha)\}}.$$

Recall that $1 \leqslant c_1(\alpha) \leqslant c_2(\alpha)$, so on the set $\{Y_T \geqslant c_2(\alpha)\}$,

$$(S_T^1 - S_T^2)^+ \mathbb{1}_{\{Y_T \geqslant c_2(\alpha)\}} = (S_T^1 - S_T^2)\mathbb{1}_{\{Y_T \geqslant c_2(\alpha)\}}.$$

These equalities and previous considerations give us

$$(S_T^1 - S_T^2)^+ \mathbb{1}_{\{Y_T \leqslant c_1(\alpha)\} \cup \{Y_T \geqslant c_2(\alpha)\}}$$
$$= (S_T^1 - S_T^2)^+ - S_T^1 + S_T^2 + (S_T^1 - S_T^2)\mathbb{1}_{\{Y_T \leqslant c_1(\alpha)\}} + (S_T^1 - S_T^2)\mathbb{1}_{\{Y_T \geqslant c_2(\alpha)\}}$$
$$= (S_T^1 - S_T^2)^+ + (S_T^1 - S_T^2)\mathbb{1}_{\{Y_T \leqslant c_1(\alpha)\}} - (S_T^1 - S_T^2)\mathbb{1}_{\{Y_T \leqslant c_2(\alpha)\}},$$

as

$$\mathbb{1}_{\{Y_T \geqslant c_2(\alpha)\}} = 1 - \mathbb{1}_{\{Y_T < c_2(\alpha)\}} = 1 - \mathbb{1}_{\{Y_T \leqslant c_2(\alpha)\}}.$$

With the help of the above equation, we calculate the price of our contract for case 3:

$$\begin{aligned} C &= C^{\text{Mar}} + S_0^1 \Phi(-b_+(c_1)) - S_0^2 \Phi(-b_-(c_1)) \\ &\quad - S_0^1 \Phi(-b_+(c_2)) + S_0^2 \Phi(-b_-(c_2)) \\ &= S_0^1 \left[\Phi(b_+(1)) - \Phi(b_+(c_1)) + \Phi(b_+(c_2)) \right] \\ &\quad - S_0^2 \left[\Phi(b_-(1)) - \Phi(b_-(c_1)) - \Phi(b_-(c_2)) \right]. \end{aligned} \quad (2.77)$$

Similarly, we find the survival probability for the third case:

$$_T p_x = 1 - \frac{S_0^1 \left[\Phi(b_+(c_1)) - \Phi(b_+(c_2)) \right] - S_0^2 \left[\Phi(b_-(c_1)) - \Phi(b_-(c_2)) \right]}{S_0^1 \Phi(b_+(1)) - S_0^2 \Phi(b_-(1))}. \quad (2.78)$$

As a result, we arrive at the following theorem.

Theorem 2.2. *Assume the model (2.39) satisfies at (2.67), and $\rho^2 < 1$. Then the quantile prices of the contract with its embedded call option (2.54)–(2.55) and survival probabilities are determined by the formulas (2.74)–(2.78).*

We will calculate now the capital for the optimal hedging strategy π^* and find the values of its components $(\beta_t^*, \gamma_t^{1*}, \gamma_t^{2*})$, the amounts that should be invested in the bank account and the two assets.

From general theory, we know that the capital for the perfect hedging strategy π^* at any time is the expectation of the contingent claim f_T, conditioned on the information until the present moment t:

$$\frac{X_t^{\pi^*}}{e^{rt}} = \mathbb{E}^* \left(\frac{f_T}{e^{rT}} \bigg| \mathcal{F}_t \right) \Leftrightarrow X_t^{\pi^*} = \mathbb{E}^* \left(\frac{f_T}{e^{r(T-t)}} \bigg| \mathcal{F}_t \right).$$

Recall that π^* is the optimal hedge for $(S_T^1 - S_T^2)^+$ under the restriction of limited initial capital x, insufficient for the perfect hedge of this payoff. From quantile hedging theory (see (2.49)), we know that π^* becomes the perfect hedge for the modified claim

$$(S_T^1 - S_T^2)^+ \mathbb{1}_{A^*}, \quad (2.79)$$

so our task is to calculate

$$X_t^{\pi^*} = \mathbb{E}^* \left(\frac{(S_T^1 - S_T^2)^+}{e^{r(T-t)}} \mathbb{1}_{A^*} \bigg| \mathcal{F}_t \right).$$

Now, in the previous theorem we showed that the maximal set of successful hedging A^* could be represented in terms of $Y_T = \frac{S_T^1}{S_T^2}$ (2.68), with three cases as solutions to the equation (2.69). We found the price of the contingent claim separately for the different possibilities. Here we use the same approach to calculate $X_t^{\pi^*}$.

For cases 1 and 2, with $-\alpha < 0$ and $0 \leqslant -\alpha \leqslant 1$, we have

$$E^*\left(\frac{(S_T^1 - S_T^2)^+}{e^{r(T-t)}}\mathbb{1}_{A^*}\bigg|\mathcal{F}_t\right) = E^*\left(\frac{(S_T^1 - S_T^2)^+}{e^{r(T-t)}}\mathbb{1}_{\{Y_T \leqslant c(\alpha)\}}\bigg|\mathcal{F}_t\right)$$

$$= E^*\left(\frac{(S_T^1 - S_T^2)^+}{e^{r(T-t)}}\bigg|\mathcal{F}_t\right) - E^*\left(\frac{S_T^1}{e^{r(T-t)}}\bigg|\mathcal{F}_t\right) + E^*\left(\frac{S_T^2}{e^{r(T-t)}}\bigg|\mathcal{F}_t\right) \quad (2.80)$$

$$+ E^*\left(\frac{(S_T^1 - S_T^2)}{e^{r(T-t)}}\mathbb{1}_{\{Y_T \leqslant c(\alpha)\}}\bigg|\mathcal{F}_t\right).$$

The first term after the equal sign in the above formula is determined by Margrabe's formula for the capital of the perfect hedge at any time t:

$$E^*\left(e^{-r(T-t)}(S_T^1 - S_T^2)^+\big|\mathcal{F}_t\right) = \mathcal{C}^{\mathrm{Mar}}(S_t^1, S_t^2, T-t)$$
$$= S_t^1 \Phi(b_+(T-t)) - S_t^2 \Phi(b_-(T-t)),$$

$$b_\pm(S_t^1/S_t^2, T-t) = \frac{\ln\frac{S_0^1}{S_0^2} \pm \frac{\sigma^2}{2}(T-t)}{\sigma\sqrt{T-t}}. \quad (2.81)$$

Simplifying (2.80), we get

$$E^*\left(\frac{(S_T^1 - S_T^2)^+}{e^{r(T-t)}}\mathbb{1}_{A^*}\bigg|\mathcal{F}_t\right) = \mathcal{C}^{\mathrm{Mar}}(S_t^1, S_t^2, T-t) - S_t^1 + S_t^2$$

$$+ S_t^1 E^*\left(\frac{S_{T-t}^1}{e^{r(T-t)}}\mathbb{1}_{\{Y_{T-t} \leqslant c(\alpha)/Y_t\}}\bigg|\mathcal{F}_t\right) \quad (2.82)$$

$$- S_t^2 E^*\left(\frac{S_{T-t}^2}{e^{r(T-t)}}\mathbb{1}_{\{Y_{T-t} \leqslant c(\alpha)/Y_t\}}\bigg|\mathcal{F}_t\right),$$

where the notation S_{T-t}^i, Y_{T-t} is used for $\frac{S_T^i}{S_t^i}$ and $\frac{Y_T}{Y_t}$, respectively.

Note that both assets S_T^i are based on Wiener processes W_T^i, whose increments $W_T^i - W_t^i$ depend on their last value W_t^i only (they are independent of the information before t). Also, at time t, the values S_t^1 and Y_t become deterministic. All this allows us to proceed with calculations in a straightforward manner, applying the Lemma 2.3 to find the unknowns in (2.82).

So now we have

$$\zeta = \ln(Y_{T-t}) = -\frac{\sigma_1^2 - \sigma_2^2}{2}(T-t) + \sigma(W_T^* - W_t^*)$$

$$\eta_i = -r(T-t) + \frac{\sigma_i^2}{2}(T-t) - \sigma_i(W_T^{i*} - W_t^{i*}),$$

with parameters

$$\mu_\zeta = -\frac{\sigma_1^2 - \sigma_2^2}{2}(T-t),$$

$$\sigma_\zeta = \sigma^2(T-t);$$

$$\mu_{\eta_i} = -r(T-t) + \frac{\sigma_i^2}{2}(T-t),$$

$$\sigma_{\eta_i} = \sigma_i^2(T-t);$$

$$\text{cov}((W_T^* - W_t^*), (W_T^{1*} - W_t^{1*})) = -\sigma_1(\sigma_1 - \sigma_2\rho)(T-t),$$

$$\text{cov}((W_T^* - W_t^*), (W_T^{2*} - W_t^{2*})) = -\sigma_2(\sigma_1\rho - \sigma_2)(T-t).$$

From these, we find

$$S_t^1 \mathrm{E}^*\left(\frac{S_{T-t}^1}{e^{r(T-t)}} 1_{\{Y_{T-t} \leqslant c(\alpha)/Y_t\}} \middle| \mathcal{F}_t\right) = S_t^1 \Phi\left(-\left[\frac{\ln\frac{Y_t}{c(\alpha)} + \frac{1}{2}\sigma^2(T-t)}{\sigma\sqrt{T-t}}\right]\right)$$

$$= S_t^1 \Phi(-b_+(Y_t/c(\alpha), T-t));$$

$$S_t^2 \mathrm{E}^*\left(\frac{S_{T-t}^2}{e^{r(T-t)}} 1_{\{Y_{T-t} \leqslant c(\alpha)/Y_t\}} \middle| \mathcal{F}_t\right) = S_t^2 \Phi\left(-\left[\frac{\ln\frac{Y_t}{c(\alpha)} - \frac{1}{2}\sigma^2(T-t)}{\sigma\sqrt{T-t}}\right]\right)$$

$$= S_t^1 \Phi(-b_-(Y_t/c(\alpha), T-t)).$$

Finally, taking into account the above calculations, we derive the formula for the capital at time t for cases 1 and 2:

$$\begin{aligned}
X_t^{\pi^*} &= \mathrm{E}^*\left(\frac{(S_T^1 - S_T^2)^+}{e^{r(T-t)}} 1_{A^*} \middle| \mathcal{F}_t\right) \\
&= \mathcal{C}^{\mathrm{Mar}}(S_t^1, S_t^2, T-t) - S_t^1 + S_t^2 \\
&\quad + S_t^1 \Phi\left(-b_+\left(\frac{Y_t}{c}, T-t\right)\right) - S_t^2 \Phi\left(b_-\left(\frac{Y_t}{c}, T-t\right)\right) \\
&= S_t^1 \left[\Phi\left(b_+\left(\frac{S_t^1}{S_t^2}, T-t\right)\right) - \Phi\left(b_+\left(\frac{S_t^1}{S_t^2 \cdot c}, T-t\right)\right)\right] \\
&\quad - S_t^2 \left[\Phi\left(b_-\left(\frac{S_t^1}{S_t^2}, T-t\right)\right) - \Phi\left(b_-\left(\frac{S_t^1}{S_t^2 \cdot c}, T-t\right)\right)\right],
\end{aligned} \quad (2.83)$$

with $b_\pm$ given by (2.81).

For case 3, we apply a similar analysis (see also calculations for (2.77)) and find that

$$\begin{aligned}
X_t^{\pi^*} &= \mathrm{E}^*\left(\frac{(S_T^1 - S_T^2)^+}{e^{r(T-t)}} 1_{\{Y_T \leqslant c_1(\alpha)\} \cup \{Y_T \geqslant c_2(\alpha)\}} \middle| \mathcal{F}_t\right) \\
&= S_t^1 \left[\Phi\left(b_+\left(\frac{S_t^1}{S_t^2}, T-t\right)\right) - \Phi\left(b_+\left(\frac{S_t^1}{S_t^2 \cdot c_1}, T-t\right)\right)\right. \\
&\quad \left. + \Phi\left(b_+\left(\frac{S_t^1}{S_t^2 \cdot c_2}, T-t\right)\right)\right] \\
&\quad - S_t^2 \left[\Phi\left(b_-\left(\frac{S_t^1}{S_t^2}, T-t\right)\right) - \Phi\left(b_-\left(\frac{S_t^1}{S_t^2 \cdot c_1}, T-t\right)\right)\right. \\
&\quad \left. + \Phi\left(b_-\left(\frac{S_t^1}{S_t^2 \cdot c_2}, T-t\right)\right)\right].
\end{aligned} \quad (2.84)$$

From (2.40)–(2.41) and equalities (2.83)–(2.84), we conclude that in the first two cases, the components of the perfect hedge π^* are

$$\beta_t^* = 0,$$

$$\gamma_t^{1*} = \left[\Phi\left(b_+\left(\frac{S_t^1}{S_t^2}, T-t\right)\right) - \Phi\left(b_+\left(\frac{S_t^1}{S_t^2 \cdot c}, T-t\right)\right)\right], \quad (2.85)$$

$$\gamma_t^{2*} = -\left[\Phi\left(b_-\left(\frac{S_t^1}{S_t^2}, T-t\right)\right) - \Phi\left(b_-\left(\frac{S_t^1}{S_t^2 \cdot c}, T-t\right)\right)\right].$$

For case 3, the strategy becomes

$$\beta_t^* = 0,$$

$$\gamma_t^{1*} = \left[\Phi\left(b_+\left(\frac{S_t^1}{S_t^2}, T-t\right)\right) - \Phi\left(b_+\left(\frac{S_t^1}{S_t^2 \cdot c_1}, T-t\right)\right)\right.$$

$$\left. + \Phi\left(b_+\left(\frac{S_t^1}{S_t^2 \cdot c_2}, T-t\right)\right)\right], \quad (2.86)$$

$$\gamma_t^{2*} = -\left[\Phi\left(b_-\left(\frac{S_t^1}{S_t^2}, T-t\right)\right) - \Phi\left(b_-\left(\frac{S_t^1}{S_t^2 \cdot c_1}, T-t\right)\right)\right.$$

$$\left. + \Phi\left(b_-\left(\frac{S_t^1}{S_t^2 \cdot c_2}, T-t\right)\right)\right].$$

Therefore, the investor who wishes to replicate the contingent claim (2.79) would apply the formulas (2.85)–(2.86). This strategy will maximize the probability of hedging the given equity-linked life insurance contract with a stochastic/flexible guarantee, with the payoff conditioned on the survival of the client to maturity.

2.3 Quantile portfolio management with multiple contracts with a numerical example

Suppose that an insurance firm wants to market the type of equity-linked life insurance contract discussed in Section 2.2. Then of course, the company has to hedge their short position. As we know, quantile hedging maximizes the probability of a successful hedge, that is, $P(A^*)$ is maximized (see (2.47)). However, this information may not be very useful to the hedger; instead, he/she would rather know what is the actual probability that the firm will default (fail to hedge successfully). So let us look at the problem of optimal pricing and hedging from a different perspective: can the company fix the desired level of financial risk ϵ and then find the optimal price and hedge? The answer to this question is yes; the probability of successful hedging will be

$$P(A^*) = 1 - \epsilon, \quad (2.87)$$

Quantile hedging of equity-linked life insurance contracts   47

where A^* is the maximal set of successful hedging for contingent claim H.

Once the acceptable financial risk level ϵ (the probability that the hedge will fail) is set, we can find the unknown constant c in pricing and hedging formulas (see, for instance, (2.74),(2.76),(2.77)). We are able to do this, because A^* can be rewritten in terms of the values of two risky assets at maturity and c. This further allows us to calculate the fair price and optimal hedging strategy for the given risk level.

On the other hand, relation (2.57) enables the hedger to set its acceptable level of insurance risk by targeting clients in certain age groups. In this approach, the survival probability $_Tp_x$, which reflects insurance risk, is chosen first; then, the initial price and probability of successful hedging are calculated (note that this probability will be the maximal probability of successful hedging, but it may or may not be acceptable to the hedger).

We mentioned that (2.57) allows the option writer to balance between financial and insurance risk levels. The idea is the following: the hedger chooses the acceptable financial risk level ϵ, from which we can calculate the initial price for the equity-linked insurance contract using (2.87). Then we calculate survival probabilities (from (2.57)), and, using life tables, figure out which age groups should be targeted for the contract. If the company wants to market the contract to a different age group, we go backwards and figure out the probability of successful hedging. This analysis is carried out until the firm is satisfied with both risk levels.

One benefit of this approach is obvious: the hedging firm has control over the choice of financial risk or desirable customers, while being aware of the resulting risk levels at all times. But there is also a clear drawback: the two risk components are linked and dependent on each other. However, it turns out that if the company owns a portfolio of N options, it is possible to separate financial risk from the cumulative insurance risk and to set their levels independently. We illustrate this idea below.

Let us consider a more realistic situation: suppose the insurance company wants to offer the same contract to a (homogeneous) group of size l_x of clients of age x. Then the firm will have a cumulative liability $l_x \cdot H$ at maturity. So, initially, the company just charges each individual the fair price $\mathcal{C}$ (as determined above) and invests the amount $l_x \cdot \mathcal{C}$ for the optimal hedge of the full amount $l_x \cdot H$.

However, this may not be the best decision on the part of the risk managers, as some of the clients in a large group may not live to maturity of the option to collect the payoff, and the company will have overhedged. Below, we explain how to take this issue into account.

We assume that the clients' remaining lifetimes $\{T_i(x)\}$, $i = 1, 2, \cdots, l_x$, are independent identically distributed random variables on $(\widetilde{\Omega}, \widetilde{\mathcal{F}}, \widetilde{\mathbb{P}})$, the probability space reflecting insurance risk. Then the payoff becomes

$$(S_T^1 - S_T^2)\mathbb{1}_{\{T_i(x) > T\}}. \qquad (2.88)$$

Clearly, the indicator function above takes value 1 with probability $_T p_x$ and 0 with probability $1 - {_T p_x}$ (Bernoulli random variable).

Consider that d_T^x defines the number of deaths that have occurred in the group of insured clients by time $t \leqslant T$. Then the number of individuals alive at time T is $l_{x+T} = l_x - d_T^x$, and l_{x+T} is a binomial random variable with probability of success $_T p_x$.

So the insurance company can take into account the evolution of the entire group and hedge only the amount $l_{x+T} \cdot H$. But we wish to replace l_{x+T} with some constant n, otherwise the pricing and hedging calculations get significantly more complex. To do this, the insurance company fixes some level of cumulative insurance risk δ and finds $n = n_\delta$ from the condition

$$\widetilde{P}\{l_{x+T} \leqslant n_\delta\} \geqslant 1 - \delta, \quad \delta \in (0,1). \tag{2.89}$$

Now only $n_\delta \cdot H$, not the full amount $l_x \cdot H$, must be hedged. This lowers the initial price of the contract for every individual in the group and, as a results, attracts more clients.

Thus at maturity the firm's liability will equal $\frac{n_\delta}{l_x}(S_T^1 - S_T^2)^+$, with the constant n_δ previously determined from the level of insurance risk.

Recall that when applying quantile hedging, we end up hedging perfectly the modified payoff H^*. It turns out that we can rewrite $H^* = H \cdot \mathbb{1}_{A^*}$ for quantile hedging.

Now, since A^* is the maximal set of successful hedging, in cases of multiple contracts,

$$A^* = \left\{l_x \cdot X_T^{\pi^*} \geqslant n_\delta (S_T^1 - S_T^2)^+\right\} = \left\{X_T^{\pi^*} \geqslant \frac{n_\delta}{l_x}(S_T^1 - S_T^2)^+\right\},$$

where $X_T^{\pi^*}$ is the optimal hedging capital for the contract.

On the other hand, A^* can be represented as some set, for instance, $A^* = \{Y_T \leqslant c\}$, where $Y_T = \frac{S_T^1}{S_T^2}$. This representation is not affected by the constant $\frac{n_\delta}{l_x}$ multiplying the payoff; only the value of c will matter in pricing calculations. Thus, as one may expect, the price of each individual contract with payoff (2.88) is simply lowered by the amount $\frac{n_\delta}{l_x}$:

$$\widetilde{C} = \mathrm{E}^* \left(\frac{n_\delta}{l_x} e^{-rT}(S_T^1 - S_T^2)^+ \mathbb{1}_{A^*}\right) = \frac{n_\delta}{l_x} \cdot C, \tag{2.90}$$

where C represents the price calculated using quantile hedging, found explicitly once the financial level of risk, ϵ, is set.

To summarize, these are the steps in the suggested risk management schemes. The hedging firm decides on the level of financial risk level, ϵ. From this level, initial contract price and survival probability of the client are calculated. Using life tables, the company figures x, the appropriate age of the client for the contract. Then, the firm expands from one customer to a group of N clients of age x, collecting from each the initial price $\widetilde{C}$. Note that this price is lower than the price found using quantile hedging (paid by a single client), but the company carries higher total risk: δ, the chance that more clients than the expected n_δ will survive to T, and ϵ, the probability that the optimal hedge will be unsuccessful.

In the following we give an example of simple actuarial analysis illustrating these risk-management ideas.

Example 2.1. In this illustrative example, we used quantile hedging for pricing calculations. To find parameters in our model, we used daily stock prices of Russell-2000 (RUT-I) and Dow Jones Industrial Average (DJIA) from August 1, 1997 to July 31, 2003. The first index, RUT-I, represents 2000 smaller to companies in the US, and the second, DJIA, reflects the performance of 30 large and prestigious US companies. Since RUT-I is riskier than DJIA, we take the DJIA stock price as the flexible guarantee, S^2.

The calculated parameters for the two indices are

$$\mu_1 = .0481, \quad \sigma_1 = .2232;$$
$$\mu_2 = .0417, \quad \sigma_2 = .2089;$$
$$\rho = 0.71.$$

The condition (2.67) is approximately fulfilled, with $\alpha \approx -0.6 > -1$, so we use (2.74) and (2.75) to find initial prices and survival probabilities.

First, assume that the insurance company sells one (single contract) call option on RUT-I with the value of DJIA at maturity as the strike. The client will receive the payoff only if he/she is alive at maturity.

The initial prices of Russel-2000 and DJIA are 414.21 and 8194.04, respectively. Because of the large difference in these, we set the price of the first asset $\frac{8194.04}{414.21} \cdot S_t^1$; this makes initial values of the two assets equal. We consider three levels of financial risk, $\epsilon = .025, .05, .1$, and four maturities $T = 1, 3, 5, 10$ years.

For the payoff that is not conditioned on the client's survival to maturity, the fair price of the option is calculated using perfect hedging (2.53):

$T = 1$	$T = 3$	$T = 5$	$T = 10$
538.99	931.45	1199.77	1687.20

TABLE 2.1: Contract price if payoff is purely financial.

Now, suppose the insurance company chooses its acceptable level of financial risk ϵ. Using (2.87) and (2.74), we find the corresponding price for the contract with payoff conditioned on the client's survival to maturity (2.88) (see Table 2.2).

These results are very encouraging. A client who is interested in purchasing the call in consideration, but does not care to collect the payoff if he/she is not alive at T, will choose to go to our insurance firm and pay the lower price above rather than buying the option from a competitor for a higher price (as

	$T=1$	$T=3$	$T=5$	$T=10$
$\epsilon = 0.025$	462.75	802.32	1035.28	1459.18
$\epsilon = 0.05$	403.77	701.97	907.28	1282.01
$\epsilon = 0.1$	307.50	537.46	697.00	990.31

TABLE 2.2: Contract price calculated using quantile hedging.

in Table 2.2). Observe that the higher the risk carried by the firm, the lower the option price becomes.

Now, with the above information, the company knows the appropriate initial price and the level of financial risk inherent in the contract. But the firm has yet to figure out x, the age of clients appropriate for the option. So, we calculate survival probabilities $_T p_x$ using (2.75) and find x from life tables (in Bowers et al. (1997) see Table 2.3).

	$T=1$	$T=3$	$T=5$	$T=10$
$\epsilon = 0.025$	89	75	67	56
$\epsilon = 0.05$	97	82	75	64
$\epsilon = 0.1$	—	90	82	72

TABLE 2.3: Ages of clients

From this table, the firm can come to the following conclusions. First, since the ages are so high for $T = 1$ and $T = 3$, the company may want to consider only longer-term contracts, as it may be difficult to market three shorter-term options to clients that old. Second, the higher the financial risk carried by the firm, the higher the client ages. This result is clear: if the probability of unsuccessful hedging is greater, then it is only reasonable that this higher risk is balanced by signing contracts with older clients (as these will be less likely to survive to maturity to claim their payoffs). Third, the longer the contract, the lower the recommended ages are. Again, this is intuitive: anyone is more likely to die in 10 years than in 3 years, so the company is safer offering long-term options to younger clients.

Now, let's see how prices are affected when the firm sells a portfolio of N contracts instead of only one.

Suppose the insurance company makes the contract with $N = 100$ clients of age x, calculated previously from the level of financial risk, ϵ. Of course, in real life, the ages need not be exactly the same, just close enough, at the hedger's discretion. Next, the company chooses its acceptable level of cumulative insurance risk, δ as shown in (2.89). Consequently, the price of each individual contract is lowered by the fraction of the clients surviving to T (see (2.90)).

Let us consider the results for $x = 67$, with $T = 5$ and $\epsilon = .025$ (see Table 2.3). Out of $N = 100$, 93 individuals will survive to $T = 5$ with probability $1 - \delta = .975$. Thus the initial price paid by each client in the group is reduced from 1035.28 (Table 2.2) to $\frac{93}{100} \cdot 1035.28 = 962.81$. With this, the insurance company carries 2.5 percent financial and 2.5 percent insurance (mortality) risk.

If we take $T = 10$ and $\epsilon = .05$, then $x = 64$, and only 82 persons will survive to T with probability $1 - \delta = .95$. In this case, the individual price is reduced significantly: from 1282.01 to 1051.25. The company carries 10 percent total risk.

There are several things to point out.

First, we took financial and mortality risks to be equal, which is clearly not a limitation.

Second, since survival probabilities are rather high for all age groups for short periods of time, almost all clients in the group will survive to a contract's maturity (for $T = 1, 3$, the resulting price reductions are only a few dollars, so we omit these calculations here).

Thus the hedger's risk-management approach will be similar for both a single option and a portfolio of N options, provided that these are short-term.

We start observing significant price reductions for constants of longer duration, so the insurance company clearly benefits from selling many of these: because the price for each individual is lowered greatly, the firm has an advantage when competing for clients with other companies.

Chapter 3

Valuation of equity-linked life insurance contracts via efficient hedging in the Black–Scholes model

3.1   Efficient hedging and pricing formulas for contracts with
      correlated assets and guarantees 53
3.2   Efficient hedging and pricing of contracts with perfectly
      correlated stocks and guarantees 64
3.3   Illustrative risk-management for the risk-taking insurer 72

3.1 Efficient hedging and pricing formulas for contracts with correlated assets and guarantees

Here we develop an efficient hedging methodology to calculate theoretical price and survival probabilities for equity-linked life insurance contracts with stochastic/flexible guaranties. We work in the model given in details in Section 2.2, formulas (2.39)–(2.45). The corresponding description of the contract under consideration is presented by formulas (2.50)–(2.52) and (2.54)–(2.55). In contrast with the problem of Section 2.2, the writer of the contract with maturity time T faces a possibility of a loss, created by the difference between the contract's payoff and the value of the capital generated by the initially invested amount $(H - X_T)^+$, which is called a shortfall.

To put forward the efficient hedging problem, we introduce a loss function $l = l(x)$, $x \geq 0$, with $l(0) = 0$ and choose the work with a power loss function

$$l(x) = \text{const} \cdot x^p, \quad p > 0, \; x \geq 0. \tag{3.1}$$

We will conduct analysis for three possible cases of p: $p = 1$, $p > 1$ and $0 < p < 1$.

Define the shortfall risk as $\mathrm{E}\left(l(H - X_T)^+\right)$. The goal of efficient hedging is finding an optimal strategy π^* that minimizes the shortfall risk under the restriction on the initial capital x of admissible strategy π

$$\mathrm{E}\left(l(H - X_T^{\pi*})^+\right) = \inf_{\pi} \mathrm{E}\left(l(H - X_T^{\pi})^+\right), \tag{3.2}$$

where $x = X_0^{\pi} \leq X_0 < \mathrm{E}^* H e^{-rT}$.

The strategy π^* is called the efficient hedge. Once the efficient hedge is constructed, we will set the initial price of the contract as equal to its initial values $X_0^{\pi*}$.

Due to Föllmer and Leukert the problem (3.2) can be reformulated in the following way:
find an $\mathcal{F}_t$-measurable function (decision rule) $\varphi : \Omega \to [0,1]$ to minimize

$$\mathrm{E}\left(l(1-\varphi)H\right) \text{ under constrain } \mathrm{E}^*\varphi H \leqslant X_0. \tag{3.3}$$

The problem (3.3) and, correspondingly, (3.2) can be solved with the help of the fundamental Neyman-Pearson lemma, and we formulate this result of Föllmer and Leukert as the following.

Lemma 3.1. *Consider a contingent claim with the payoff H at maturity with the shortfall from its hedging weighted by a power loss function (3.1). Then the efficient hedge π^* satisfying (3.2) exists and coincides with a perfect hedge for a modified claim H_p having the structure:*

$$H_p = H - a_p Z_T^{\frac{1}{p-1}} \wedge H \text{ for } p > 1, \text{ const } = \frac{1}{p},$$
$$H_p = H \mathbb{1}_{\{Z_t^{-1} > a_p H^{1-p}\}} \text{ for } 0 < p < 1, \text{ const } = 1, \tag{3.4}$$
$$H_p = H \mathbb{1}_{\{Z_t^{-1} > a_p\}} \text{ for } p = 1, \text{ const } = 1,$$

where a constant a_p is defined from the condition on its initial value $\mathrm{E}^* H_p e^{-rT} = X_0$, Z_T is defined by (2.43).

The power loss function is one example of an interesting loss function: its concavity indicates risk preferences of the hedge and varies with different values of p.

When $p > 1$, l is concave up, the investor is risk-averse: for the difference $(H - X_T)$ of size ϵ, the expected loss is greater than ϵ. The investor's goal is to make this larger potential loss as small as possible.

Compare to the case when $0 < p < 1$: here l is concave down, and the expected loss is smaller than ϵ, so minimizing it holds less value than in the previous scenario, meaning that now the investor is a risk-taker.

When $p = 1$, the investor is indifferent between risk-aversion and risk-taking.

Due to these varied behaviors of power loss functions, we analyze the three possibilities for p separately and, using these results, find the solution structure in each case.

Let us look at a contract where the payoff $H = (S_T^1 - S_T^2)^+$, but is now conditioned on the client's mortality. This is a type of knock-in option: the payoff only happens if it is "knocked-in" by the client's survival to the expiry date T. Equivalently, the death of the insured would knock-out the contract: if the insured person dies before T, the contract ceases to exist. Clearly we have two sources of risk here, the volatility of the risky assets (financial risk)

Valuation of equity-linked life insurance contracts 55

and the mortality of the client (insurance risk). Formally, we represent the payoff of such a mixed contract as

$$(S_T^1 - S_T^2)^+ \mathbb{1}_{\{T(x)>T\}}. \tag{3.5}$$

Recall that $T(x)$ is the remaining life of the time insured of age x (see (2.50)).

From general theory, we know that the initial price of a contract is the expectation w.r. to risk-neutral measure (P^* in our case) of the contract's discounted payoff. Taking into account (2.50), we calculate the Brennan–Schwartz initial price $_TU_x$ of our contract (3.5):

$$_TU_x = \mathrm{E}^* \times \widetilde{\mathrm{E}}\left[e^{-rT}(S_T^1 - S_T^2)^+ \mathbb{1}_{\{T(x)>T\}}\right] = \mathrm{E}^*\left[e^{-rT}(S_T^1 - S_T^2)^+\right] \cdot {_Tp_x}.$$

Previously we mentioned that the mortality of the client (represented by $_Tp_x$) causes the price of the contract to decrease:

$$(S_T^1 - S_T^2)^+ \cdot {_Tp_x} < (S_T^1 - S_T^2)^+ \text{ implies that}$$
$$e^{-rT}\mathrm{E}^*\left[(S_T^1 - S_T^2)^+\right] \cdot {_Tp_x} < e^{-rT}\mathrm{E}^*\left[(S_T^1 - S_T^2)^+\right] = X_0.$$

At the price paid by the client is insufficient for a perfect hedge, the insurance company can use an efficient hedging methodology to minimize the shortfall risk.

Combining general pricing theory and efficient hedging (see Lemma 3.1), we observe that

$$_TU_x = \mathrm{E}^*\left[e^{-rT}(S_T^1 - S_T^2)^+\right] \cdot {_Tp_x} = \mathrm{E}^*\left[e^{-rT}(S_T^1 - S_T^2)_p^+\right]. \tag{3.6}$$

Equality (3.6) enables us to derive the following expression for the survival probability $_Tp_x$:

$$_Tp_x = \frac{\mathrm{E}^*\left[(S_T^1 - S_T^2)_p^+\right]}{\mathrm{E}^*\left[e^{-rT}(S_T^1 - S_T^2)^+\right]}, \tag{3.7}$$

where the modified payoff $(S_T^1 - S_T^2)_p^+$ depends on the value of p of the power loss function. The equation (3.7) is essential to our actuarial analysis, because it quantifies the relationship between the insurance and the financial components of the contract's risk. Moreover, (3.7) shows that if an insurance company takes on too much financial risk, it can compensate by allowing less insurance risk (signing contracts with older clients, for instance), and vice versa. The reader will have an opportunity to see a numerical illustration of this type of actuarial analysis at the end of this section.

There is another important point to emphasize. Previously we discussed only the minimization of the shortfall risk given some initial capital. However, if the company is willing to take on some risk in return for offering cheaper contracts to attract more clients, we can calculate the initial price that should be charged from the level of risk the company is willing to undertake. But we can do even more: using (3.7), we can also recover the recommended ages of clients with whom the contracts should be made.

Before we move on to price calculations, we note here that

$$W_t^* = \frac{\sigma_1 W_t^{1*} - \sigma_2 W_t^{2*}}{\sigma}$$

is a new Brownian motion w.r. to P*, with σ defined in (2.53), and with covariances given by

$$\mathrm{cov}(W_t^*, W_t^{1*}) = \mathrm{E}^*(W_t^* \cdot W_t^{1*}) = \frac{(\sigma_1 - \sigma_2 \rho)t}{\sigma},$$

$$\mathrm{cov}(W_t^*, W_t^{2*}) = \mathrm{E}^*(W_t^* \cdot W_t^{2*}) = \frac{(\sigma_1 \rho - \sigma_2)t}{\sigma}.$$

Now, we will also need to work with density $Z_T = \frac{dP_T^*}{dP_T}$ as a function of both risky assets. After some manipulations, we get this expression for Z in terms of S^1 and S^2:

$$Z_T = \exp\left\{\varphi_1 W_T^1 + \varphi_2 W_T^2 - \frac{1}{2}\left(\varphi_1^2 + \varphi_2^2 + 2\rho\varphi_1\varphi_2\right)T\right\}$$

$$= (S_T^1)^{\frac{\varphi_1}{\sigma_1}} \cdot (S_T^2)^{\frac{\varphi_2}{\sigma_2}} \cdot g, \tag{3.8}$$

where

$$g = \frac{\exp\left\{\frac{\sigma_1 \varphi_1 T}{2} + \frac{\sigma_2 \varphi_2 T}{2}\right\}}{(S_0^1)^{\frac{\varphi_1}{\sigma_1}} \cdot (S_0^2)^{\frac{\varphi_2}{\sigma_2}} \cdot \exp\left\{\frac{\mu_1 \varphi_1 T}{2} + \frac{\mu_2 \varphi_2 T}{2} - \frac{1}{2}\left(\varphi_1^2 + \varphi_2^2 + 2\rho\varphi_1\varphi_2\right)T\right\}}.$$

Now let us analyze the three cases for our loss function separately.
Case 1: $p > 1$
To make calculations easier, we want to rewrite H_p in terms of $Y_T = \frac{S_T^1}{S_T^2}$. In this particular case, taking into account (3.8), we represent H_p in (3.4) as follows:

$$H_p = H - \mathrm{const} \cdot Z_T^{\frac{1}{p-1}} \wedge H$$

$$= (S_T^1 - S_T^2)^+ - \mathrm{const} \cdot Z_T^{\frac{1}{p-1}} \wedge (S_T^1 - S_T^2)^+$$

$$= \left((S_T^1 - S_T^2)^+ - \mathrm{const} \cdot Z_T^{\frac{1}{p-1}}\right) \cdot \mathbb{1}_{\left\{\mathrm{const} \cdot Z_T^{\frac{1}{p-1}} < (S_T^1 - S_T^2)^+\right\}} \tag{3.9}$$

$$= \left((S_T^1 - S_T^2)^+ - \mathrm{const} \cdot Z_T^{\frac{1}{p-1}}\right) \cdot \mathbb{1}_{\left\{\frac{\mathrm{const} \cdot Z_T^{\frac{1}{p-1}}}{S_T^2} < (Y_T - 1)^+\right\}}$$

$$= \left((S_T^1 - S_T^2)^+ - \mathrm{const} \cdot Z_T^{\frac{1}{p-1}}\right) \cdot \mathbb{1}_{\left\{Y_T^\alpha \cdot g^{\frac{1}{p-1}} \cdot \mathrm{const} < (Y_T - 1)^+\right\}}.$$

From the set in the last line, we have a condition for our model:

$$\alpha = \alpha_p = \frac{\varphi_1}{\sigma_1 \cdot (p-1)} = -\left(\frac{\varphi_2}{\sigma_2 \cdot (p-1)} - 1\right). \quad (3.10)$$

Consider the characteristic equation

$$\text{const} \cdot g^{\frac{1}{p-1}} \cdot y^\alpha = (y-1)^+. \quad (3.11)$$

If $\alpha \leq 1$, we have one solution c to (3.11), and the last set in (3.9) becomes $\{Y_T \geq c\}$. For $\alpha > 1$, there two solutions, c_1 and c_2, and we obtain $\{Y_T < c_1\} \cup \{Y_T > c_2\}$ as our solution set. Note that $c \geq 1$ and $c_2 \geq c_1 \geq 1$.

First, consider the case $\alpha \leq 1$. We further rewrite H_p as

$$H_p = \left((S_T^1 - S_T^2)^+ - \text{const} \cdot Z_T^{\frac{1}{p-1}}\right) \cdot \mathbb{1}_{\{Y_T \geq c\}}$$

$$= \left((S_T^1 - S_T^2)^+ - \text{const} \cdot g^{\frac{1}{p-1}} \cdot Y_T^\alpha \cdot S_T^2\right) \cdot \mathbb{1}_{\{Y_T \geq c\}} \quad (3.12)$$

$$= (S_T^1 - S_T^2) - (S_T^1 - S_T^2) \cdot \mathbb{1}_{\{Y_T \leq c\}} - \frac{c-1}{c^\alpha} \cdot Y_T^\alpha \cdot S_T^2 \cdot \mathbb{1}_{\{Y_T \geq c\}}.$$

(We solved (3.12) to find the constant in the last line above.)

Now, taking expectations w.r. to the risk-neutral measure P*, we find from (3.12) that

$$E^*\left(\frac{H_p}{e^{rT}}\right) = S_0^1 - S_0^2 - S_0^1 \Phi(-b_+(c)) + S_0^2 \Phi(-b_-(c))$$

$$- \frac{c-1}{c^\alpha} \left[E^*\left(\frac{Y_T^\alpha \cdot S_T^2}{e^{rT}}\right) - E^*\left(\frac{Y_T^\alpha \cdot S_T^2}{e^{rT}} \mathbb{1}_{\{Y_T \leq c\}}\right)\right]. \quad (3.13)$$

Introducing some new constants, we can express $Y_T^\alpha \cdot S_T^2$ as follows:

$$Y_T^\alpha \cdot S_T^2 = \frac{S_T^{1\alpha}}{S_T^{2\alpha-1}} = h \cdot e^{\left\{\sigma_\alpha W_T^\alpha + \left(r - \frac{\sigma_\alpha^2}{2}\right)T\right\}}, \quad (3.14)$$

where

$$h = (S_0^1)^\alpha \cdot (S_0^2)^{1-\alpha} \cdot e^{\left\{\left(\frac{\sigma_\alpha^2 - \sigma_2^2}{2}\right)T - \left(\frac{\sigma_1^2 - \sigma_2^2}{2}\right)\alpha T\right\}},$$

$$\sigma_\alpha^2 = \sigma_1^2 \alpha^2 - 2\sigma_1 \sigma_2 \alpha(\alpha-1)\rho + \sigma_2^2(\alpha-1)^2,$$

and

$$W_T^\alpha = \frac{\sigma_1 \alpha W_T^{1*} - \sigma_2(\alpha-1)W_T^{2*}}{\sigma_\alpha}$$

is a new Brownian motion w.r. to P*.

So to find the initial price of the modified contingent claim, we calculate the last two terms in (3.13):

$$E^* \left(\frac{Y_T^\alpha \cdot S_T^2}{e^{rT}} \right) = h \cdot E^* \left(\frac{e^{\left\{ \left(r - \frac{\sigma_\alpha^2}{2}\right)T + \sigma_\alpha W_T^\alpha \right\}}}{e^{rT}} \right)$$

$$= h \cdot E^* \left(e^{\left\{ -\frac{T\sigma_\alpha^2}{2} + \sigma_\alpha W_T^\alpha \right\}} \right) = h \cdot 1 = h.$$

Now, to find $E^* \left(\frac{Y_T^\alpha \cdot S_T^2}{e^{rT}} \cdot \mathbb{1}_{\{Y_T \le c\}} \right)$, we use Lemma 2.3 (note that η and ψ are Gaussian random variables):

$$E^* \left(e^{-\eta} \cdot \mathbb{1}_{\psi \le \ln(c)} \right) = e^{\frac{\sigma_\eta^2}{2} - \mu_\eta} \cdot \Phi \left(\frac{\ln(c) - (\mu_\psi - \text{cov}(\psi, \eta))}{\sigma_\psi} \right). \qquad (3.15)$$

In our case

$$\psi = \ln Y_T = \ln \frac{S_0^1}{S_0^2} + \sigma_1 W_T^{1*} - \sigma_2 W_T^{2*} - \left(\frac{\sigma_1^2 - \sigma_2^2}{2} \right) T;$$

$$e^{-\eta} = e^{\left\{ \left(\frac{r - \sigma_\alpha^2}{2} \right) T + \sigma_\alpha W_T^\alpha \right\}}; \quad \eta = -rT + \frac{\sigma_\alpha^2}{2} T - \sigma_\alpha W_T^\alpha.$$

Calculating parameters, we find that

$$\mu_\psi = \ln \frac{S_0^1}{S_0^2} - \left(\frac{\sigma_1^2 - \sigma_2^2}{2} \right) T;$$

$$\sigma_\psi^2 = \sigma^2 T;$$

$$\mu_\eta = -rT + \frac{\sigma_\alpha^2}{2} T;$$

$$\sigma_\eta^2 = \sigma_\alpha^2 T;$$

$$\text{cov}(\eta, \psi) = (\sigma_2^2 - \sigma_1 \sigma_2 \rho - \sigma^2 \alpha) T,$$

with σ defined in (2.53).

From the above considerations, we obtain

$$E^* \left(\frac{Y_T^\alpha \cdot S_T^2}{e^{rT}} - \frac{Y_T^\alpha \cdot S_T^2}{e^{rT}} \cdot \mathbb{1}_{\{Y_T \le c\}} \right) = h \cdot \left(1 - \Phi(-(b_-(c) + \sigma \alpha \sqrt{T})) \right)$$

$$= h \cdot \Phi(b_-(c) + \sigma \alpha \sqrt{T}),$$

with $b_\pm(c)$ given in (2.53).

Finally, we put all the pieces together to derive the formula for the initial price for the case $p > 1$:

$$\mathbb{C}_p = \mathrm{E}^*\left(\frac{H_p}{e^{rT}}\right) = S_0^1 \cdot \Phi(b_+(c)) - S_0^2 \cdot \Phi(b_-(c))) \tag{3.16}$$
$$-\frac{c-1}{c^\alpha} \cdot \frac{(S_0^1)^\alpha}{(S_0^2)^{\alpha-1}} \cdot e^{\left\{\left(\frac{\sigma_\alpha^2 - \sigma_2^2}{2}\right)T - \left(\frac{\sigma_1^2 - \sigma_2^2}{2}\right)\alpha T\right\}} \cdot \Phi(b_-(c) + \sigma\alpha\sqrt{T}).$$

To avoid messy formulas, we simply note that survival probability $_T p_x$ (3.7) for this case is just the above formula, (3.16) for $\mathbb{C}_p$, divided by Margrabe's price:

$$_T p_x = \frac{\mathbb{C}_p}{S_0^1 \Phi(b_+(1)) - S_0^2 \Phi(b_-(1))}. \tag{3.17}$$

When $\alpha > 1$, we follow similar reasoning and arrive at the following expression for the price:

$$\mathbb{C}_p = \mathrm{E}^*\left(\frac{H_p}{e^{rT}}\right)$$
$$= S_0^1 \left(\Phi(b_+(c_1)) - \Phi(b_+(c_2))\right) - S_0^2 \left(\Phi(b_-(c_1)) - \Phi(b_-(c_2))\right)$$
$$+ \frac{(S_0^1)^\alpha}{(S_0^2)^{\alpha-1}} \cdot \exp\left\{\left(\frac{\sigma_\alpha^2 - \sigma_2^2}{2}\right)T - \left(\frac{\sigma_1^2 - \sigma_2^2}{2}\right)\alpha T\right\} \times \tag{3.18}$$
$$\times \left[\left(\frac{c_1 - 1}{c_1^\alpha} - \frac{c_1 - 1}{c_1^\alpha} \cdot \Phi(b_-(c_1) + \sigma\alpha\sqrt{T})\right) \right.$$
$$\left. - \left(\frac{c_2 - 1}{c_2^\alpha} - \frac{c_2 - 1}{c_2^\alpha} \cdot \Phi(b_-(c_2) + \sigma\alpha\sqrt{T})\right)\right].$$

As in (3.17), the survival probability is the above expression, (3.18) divided by $S_0^1 \Phi(b_+(1)) - S_0^2 \Phi(b_-(1))$:

$$_T p_x = \frac{\mathbb{C}_p}{S_0^1 \Phi(b_+(1)) - S_0^2 \Phi(b_-(1))}. \tag{3.19}$$

Case 2: $0 < p < 1$

Using (3.4) we rewrite H_p as

$$H_p = H \cdot \mathbb{1}_{\left\{\frac{1}{Z_T} > kH^{1-p}\right\}}$$
$$= (S_T^1 - S_T^2)^+ \cdot \mathbb{1}_{\left\{\frac{1}{Z_T} > k(S_T^1 - S_T^2)^{+\,1-p}\right\}}$$
$$= (S_T^1 - S_T^2)^+ \cdot \mathbb{1}_{\left\{\frac{1}{Z_T^{\frac{1}{p-1}}} > k^{\frac{1}{p-1}}(S_T^1 - S_T^2)^+\right\}}$$
$$= (S_T^1 - S_T^2)^+ \cdot \mathbb{1}_{\left\{\frac{Z_T^{\frac{1}{p-1}} k^{\frac{1}{p-1}}}{S_T^2} > (Y_T - 1)^+\right\}}$$
$$= (S_T^1 - S_T^2)^+ \cdot \mathbb{1}_{\left\{(k \cdot g)^{\frac{1}{p-1}} Y_T^\alpha > (Y_T - 1)^+\right\}}.$$

Here the work parameter $\alpha = \alpha_p$ is defined in the same way as in (3.10). And again, we have either one solution c to the characteristic equation

$$y^\alpha \cdot (k \cdot g)^{\frac{1}{p-1}} = (y-1)^+,$$

when $\alpha \leqslant 1$, or two, c_1 and c_2 for $\alpha > 1$. In the first case the solution set is $\{Y_T \leqslant c\}$, otherwise it becomes $\{Y_T \leqslant c_1\} \cup \{Y_T \geqslant c_2\}$, with $c_2 \geqslant c_1 \geqslant 1$.

As in the previous case, we utilize (3.15) to calculate the initial price and survival probability. For $\alpha \leqslant 1$, they are

$$\mathbb{C}_p = \mathrm{E}^* \left(\frac{H_p}{e^{rT}} \right) \tag{3.20}$$
$$= S_0^1 \left(\Phi(b_+(1)) - \Phi(b_+(c)) \right) - S_0^2 \left(\Phi(b_-(1)) - \Phi(b_-(c)) \right),$$

$$Tp_x = 1 - \frac{S_0^1 \Phi(b_+(c)) - S_0^2 \Phi(b_-(c))}{S_0^1 \Phi(b_+(1)) - S_0^2 \Phi(b_-(1))}. \tag{3.21}$$

For $\alpha > 1$, we obtain

$$\mathbb{C}_p = \mathrm{E}^* \left(\frac{H_p}{e^{rT}} \right) = S_0^1 \left(\Phi(b_+(1)) - \Phi(b_+(c_1)) + \Phi(b_+(c_2)) \right)$$
$$- S_0^2 \left(\Phi(b_-(1)) - \Phi(b_-(c_1)) + \Phi(b_-(c_2)) \right). \tag{3.22}$$

Survival probability, of course, equals the above price $\mathbb{C}_p$ in (3.22) divided by Margrabe's formula:

$$Tp_x = \frac{\mathbb{C}_p}{S_0^1 \Phi(b_+(1)) - S_0^2 \Phi(b_-(1))}. \tag{3.23}$$

Case 3: $p = 1$

Again, we are looking for a perfect hedge for the modified contingent claim H_p (see (3.4)):

$$H_p = (S_T^1 - S_T^2)^+ \cdot \mathbb{1}_{\{\frac{1}{\tilde{z}} > \tilde{c}\}}.$$

Using (3.8), we express the density as a function of Y_T, and the maximal set of successful hedging above becomes

$$\left\{ \frac{1}{\tilde{c} \cdot g} > Y_T^\alpha \right\}. \tag{3.24}$$

However, unlike in the previous two cases, here the constant α is defined differently:

$$\alpha_p = \alpha = \frac{\varphi_1}{\sigma_1} = -\frac{\varphi_2}{\sigma_2}. \tag{3.25}$$

We are solving this characteristic equation:

$$\frac{1}{g \cdot \tilde{c}} = y^\alpha.$$

This equation has one solution c. However, there are several solution sets, depending on the value of α and c. The set in (3.24) is equivalent to $\{Y_T \leq c\}$ wherever $\alpha > 0$, and to $\{Y_T \geq c\}$ when $\alpha < 0$. For $\alpha = 0$, we have that the solution set is either the entire real line, or there are no solutions at all if it happens than $\frac{1}{g \cdot \bar{c}} < 1$.

For the case $\alpha > 0$, wherever $c < 1$, $H_p = (S_T^1 - S_T^2)^+ \cdot \mathbb{1}_{\{Y_T \leq c\}} = 0$. Otherwise,
$$H_p = (S_T^1 - S_T^2)^+ \cdot \mathbb{1}_{\{Y_T \leq c\}},$$
and the price and survival probability are calculated in the same way as in (3.20) and (3.21).

Now, suppose that $\alpha < 0$. Then
$$H_p = (S_T^1 - S_T^2)^+ \cdot \mathbb{1}_{\{Y_T \geq c\}}.$$

When $c \geq 1$, this equation holds, and the formula for initial price and survival probability are

$$\mathbb{C}_p = \mathbb{E}^* \left(\frac{H_p}{e^{rT}} \right) = S_0^1 \cdot \Phi(b_+(c)) - S_0^2 \cdot \Phi(b_-(c))), \tag{3.26}$$

$$_Tp_x = 1 - \frac{S_0^1 \Phi(b_+(c)) - S_0^2 \Phi(b_-(c))}{S_0^1 \Phi(b_+(1)) - S_0^2 \Phi(b_-(1))}. \tag{3.27}$$

For $c < 1$ we have

$$H_p = (S_T^1 - S_T^2)^+ \cdot \mathbb{1}_{\{Y_T \geq c\} \cap \{Y_T \geq 1\}} + (S_T^1 - S_T^2)^+ \cdot \mathbb{1}_{\{Y_T \geq c\} \cap \{Y_T < 1\}}$$
$$= (S_T^1 - S_T^2)^+ \cdot \mathbb{1}_{\{Y_T \geq 1\}} = (S_T^1 - S_T^2)^+.$$

Taking expectations w.r. to P*, we notice that in this particular case the price turns out to be exactly Margrabe's formula (2.53). This means that the survival probability equals 1: so the insurance company can make this contract with any client it desires. But, the price it charges for the contract will have to be "full" price, the amount required for a perfect hedge.

Let us collect all the pricing formulas that have been derived before in one place. As a result we arrive at the following theorem.

Theorem 3.1. *For the market model (2.39) and for the power loss function (3.1) the initial prices of the contract (3.5) and the corresponding survival probabilities/loss ratios are determined by formulas:*

For $p > 1$ and α given by (3.10): (3.16)–(3.17) if $\alpha \leq 1$, (3.18)–(3.19) if $\alpha > 1$;

For $0 < p < 1$ and α given by (3.10): (3.20)–(3.21) if $\alpha \leq 1$, (3.22)–(3.23) if $\alpha > 1$;

For $p = 1$ and α given by (3.25): (3.20)–(3.21) if $\alpha > 0$, (3.26)–(3.27) or (2.53) and $_Tp_x = 1$ if $\alpha < 0$.

Example 3.1. Let us illustrate the results on efficient hedging. We analyze the problem by assuming that the writer of the contract (3.5), the insurance company, desires to market the agreement to three groups of customers: young (25 years old), middle-age (50 years old) and older (75 years old). The company wants to know what price it should charge each of the groups to minimize potential losses arising from the fact that whatever the amounts received are, they will be insufficient to guarantee a perfect hedge.

Note that the loss function we consider for the analysis here is $l(x) = x$, so we work with the case $p = 1$. The other two probabilities for p are more complex and do not lend to this type of analysis as well as the case $p = 1$ does. When $0 < p < 1$, we obtain a lower bound for the minimized shortfall risk, and the case $p > 1$ presents complications due to the indicated structure of the maximal set of successful hedging (see (3.4)).

For our numerical analysis, we used daily stock index prices for NASDAQ Composite (COMPQ) and Standard and Poor's 500 (SPX) from January 10, 2003, to January 9, 2004. The first, NASDAQ, is considered riskier than the S&P 500 index, so we take NASDAQ as S^1 and S&P as S^2. Note that the condition for α (3.25) is satisfied, and $\alpha \approx -20$. From the daily rate, we calculate the parameters for our model (2.39):

$$\mu_1 = .3909, \quad \mu_2 = .2046;$$
$$\sigma_1 = .2181, \quad \sigma_2 = .1653;$$
$$\rho = .9080.$$

We use $r = .10$ or 10 percent, as our interest rate. The initial prices (from January 9, 2004) of the two assets are

$$S_0^1 = 2086.92 \quad \text{and} \quad S_0^2 = 1121.859 \cdot \frac{S_0^1}{S_0^2}.$$

Note that we make the initial values of both indices equal to avoid arbitrage (and afterwards the proportion remains fixed, but the value of S^2 changes, of course).

Now, since $p = 1$ (also, $c \geq 1$), and $\alpha < 0$, we use the formulas in (2.27). First, we calculate Margrabe prices (denominator in the expression for survival probability) for maturities at 1, 3, and 5 years:

Maturity	Price
$T = 1$	80.72
$T = 3$	139.81
$T = 5$	180.81

TABLE 3.1: Margrabe prices for different maturities.

Now, from Survival Probability Tables [Bowers et al., (1997)], we look up the probability $_tp_x$ that the client of age x will live t more years. We are

Valuation of equity-linked life insurance contracts

interested in clients of ages 25, 50, and 75, and their survival probabilities for 1, 3, and 5 years:

$(t)/$ Ages of Clients (x)	25	50	75
$T = 1$	.99868	.99410	.95493
$T = 3$	.99609	.98083	.86054
$T = 5$	.99348	.96528	.76022

TABLE 3.2: Survival probabilities for different ages and maturities.

Finally, using (3.27), we calculate the numerator in the formula for survival probability, which is the appropriate initial price that the insurance company should charge for the contract (3.5):

$(t)/$ Ages of Clients (x)	25	50	75
$T = 1$	80.67	80.30	77.14
$T = 3$	139.26	137.13	120.31
$T = 5$	179.17	174.09	137.46

TABLE 3.3: Price of charges for the contract.

The above amounts, invested in the appropriate hedging strategy, will minimize the company's shortfall risk. Note that the corresponding optimal hedging strategy is the following: initially, invest $\Phi(b_+(c))$ into S^1 and $-\Phi(b_-(c))$ into S^2 (no money is invested into the bank account).

It can be shown that afterwards, for each time t, the writer of the contract should follow the same strategy, with the difference that in the expressions for $b_\pm(c)$ (2.53), T should be replaced with $T - t$, and S_0^i with S_t^i.

In conclusion, it is worthwhile to mention several things.

First, as we indicated already, the insurance company can compensate the financial market risk by undertaking less insurance risk (namely, signing contracts with older clients). And, if the company wishes to attract younger clients, it can compensate its increased insurance risk be charging them higher prices (than older ones would pay) to allow for more initial capital to be invested into a hedging strategy.

Second, in all cases the mixed contract (3.5) is cheaper than the corresponding contract which is not conditioned on the client's survival to maturity (as in (2.53)). Thus, if one cares only about receiving the payoff while alive, obviously this person would prefer the cheaper mixed agreement.

Finally, we can expand our actuarial analysis further, by setting a certain level of risk and then determining what the initial contract price should be, as well as the ages of persons with whom such contracts should be made.

3.2 Efficient hedging and pricing of contracts with perfectly correlated stocks and guarantees

We assume that our financial market here consists of a non-risky asset and two risky assets, as in Section 3.1. The first one, S^1, is more risky and may provide possible future gains. The second one, S^2, is less risky and serves as a stochastic guarantee. We investigate the case when the evolutions of two risky assets are generated by the same Wiener process. It is a natural supplement to the model of Section 3.1 with two different Wiener processes with the correlation coefficient $0 \leqslant \rho < 1$. There are at least two reasons for this.

First of all, equity-linked life insurance contracts are typically linked to traditional equities such as traded indices and mutual funds which exhibit a high positive correlation. Hence, the case $\rho = 1$ could be a reasonable and convenient approximation.

Second, the case $\rho = 1$ demands a special mathematical consideration in comparison with the case $0 \leqslant \rho < 1$ developed before.

We consider a financial market consisting of a non-risky $B_t = e^{rt}$, $r \geqslant 0$, $t \geqslant 0$, and two risky assets S^i, $i = 1, 2$, following the Black-Scholes model

$$dS_t^i = S_t^i(\mu_i dt + \sigma_i dW_t), \quad t \leqslant T, \ i = 1, 2. \tag{3.28}$$

Here μ_i and σ_i are the rate of return and the volatility of the asset S^i, $W = (W_t)_{t \leqslant T}$ is a Wiener process defined on a standard stochastic basis $(\Omega, \mathcal{F}, \mathbb{F} = (\mathcal{F}_t)_{t \leqslant T}, P)$ and T is a maturity time. We assume, for the sake of simplicity, that $r = 0$, and, therefore, $B_t = 1$ for any t. Also, we demand that $\mu_1 > \mu_2$, $\sigma_1 > \sigma_2$. These conditions are necessary since S^2 is assumed to be a flexible guarantee and, therefore, should be less risky than S^1. The initial values for both assets are supposed to be equal $S_0^1 = S_0^2 = S_0$ and are considered as initial investments in the financial market.

As we know from the Ito formula, the model (3.28) admits the following from, which is known as geometric Brownian motion

$$S_t^i = S_0^i \exp\left\{\left(\mu_i - \frac{\sigma_i^2}{2}\right) + \sigma_i W_t\right\}, \quad i = 1, 2. \tag{3.29}$$

Let us define a probability measure P* which has the following density with respect to the initial probability measure P:

$$Z_T = \exp\left\{-\frac{\mu_1}{\sigma_1} W_T - \frac{1}{2}\left(\frac{\mu_1}{\sigma_1}\right)^2\right\}. \tag{3.30}$$

Both processes, S^1 and S^2, are martingale with respect to the measure P* if the following technical condition is fulfilled:

$$\frac{\mu_1}{\sigma_1} = \frac{\mu_2}{\sigma_2}. \tag{3.31}$$

Therefore, in order to prevent the existence of arbitrage opportunities in the market we suppose that the risky assets we are working with satisfy this technical condition (3.31). Further, according to the Girsanov theorem, the process
$$W_t^* = W_t + \frac{\mu_1}{\sigma_1}t = W_t + \frac{\mu_2}{\sigma_2}t$$
is a Wiener process with respect to P*, defined by the density (3.30).

We note from (3.29) the following useful representation of the guarantee S_t^2 through the underlying risky asset S_t^1:

$$\begin{aligned}
S_t^2 &= S_0^2 \exp\left\{\sigma_2 W_t + \left(\mu_2 - \frac{\sigma_2^2}{2}\right)t\right\} \\
&= S_0 \exp\left\{\frac{\sigma_2}{\sigma_1}\left(\sigma_1 W_t + \left(\mu_1 - \frac{\sigma_1^2}{2}\right)t\right) - \frac{\sigma_2}{\sigma_1}\left(\mu_1 - \frac{\sigma_1^2}{2}\right)t + \left(\mu_2 - \frac{\sigma_2^2}{2}\right)t\right\} \\
&= (S_0)^{1-\frac{\sigma_2}{\sigma_1}}(S_1)^{\frac{\sigma_2}{\sigma_1}} \exp\left\{-\frac{\sigma_2}{\sigma_1}\left(\mu_1 - \frac{\sigma_1^2}{2}\right)t + \left(\mu_2 - \frac{\sigma_2^2}{2}\right)t\right\}.
\end{aligned}$$
(3.32)

This shows that our setting is equivalent to one with a financial market consisting of a single risky asset and a stochastic guarantee being the function of a price of this asset.

As in Section 2.2, we will call any process $\pi_t = (\beta_t, \gamma_t^1, \gamma_t^2)_{t \geq 0}$, adapted to the price evolution $\mathcal{F}_t$, a strategy/portfolio. Define its value as a sum $X_t^\pi = \beta_t + \gamma_t^1 S_t^1 + \gamma_t^2 S_t^2$. We shall consider only self-financing strategies satisfying the following condition

$$dX_t^\pi = \beta_t + \gamma_t^1 dS_t^1 + \gamma_t^2 dS_t^2,$$

where all stochastic differentials are well defined. Every $\mathbb{F}_T$-measurable non-negative random variable H is called a contingent claim.

Following actuarial tradition, we use a random variable $T(x)$ on a probability space $(\widetilde{\Omega}, \widetilde{\mathcal{F}}, \widetilde{\mathbb{P}})$ to denote the remaining lifetime of a person of age x. Let $_T p_x = \widetilde{\mathbb{P}}\{T(x) > T\}$ be a survival probability for the next T years of the same insured. It is reasonable to assume that $T(x)$ does not depend on the evolution of the financial market.

We study pure endowment contracts with a flexible stochastic guarantee which make a payment at maturity provided an insured is alive. Due to independency of "financial" and "insurance" parts of the contract, we consider the product probability space $(\Omega \times \widetilde{\Omega}, \Omega \times \widetilde{\mathcal{F}}, \mathbb{P} \times \widetilde{\mathbb{P}})$ and introduce a contingent claim as (2.54)

$$H(T(x)) = \max\{S_T^1, S_T^2\} \cdot \mathbb{1}_{\{T(x) > T\}}. \tag{3.33}$$

It is known that a strategy with the payoff $H = \max\{S_T^1, S_T^2\}$ at T is a perfect hedge for the contract, and its price is equal to $\mathbb{E}^* H$.

Let us rewrite the financial component of (3.33) as follows:

$$H(T(x)) = \max\{S_T^1, S_T^2\} = S_T^2 + (S_T^1 - S_T^2)^+. \tag{3.34}$$

Using (3.34) we reduce the pricing of the claim (3.33) to the pricing of the call option $(S_T^1 - S_T^2)^+$ provided $\{T(x) > T\}$.

According to the well-developed option pricing theory, the optimal price is traditionally calculated as an expected present value of cash flows under the risk-neutral probability measure. Note, however, that the "insurance" part of the contract (3.33) does not need to be risk-adjusted since the mortality risk is essentially unsystematic. It means that the mortality risk can be effectively eliminated by increasing the number of similar insurance policies. These ideas lead to the so-called Brennan–Schwartz price of (3.33)

$$_T U_x = \text{E}^* \times \tilde{\text{E}} H(T(x)) = {_T p_x} \text{E}^*(S_T^2) + {_T p_x} \text{E}^*(S_T^1 - S_T^2)^+, \quad (3.35)$$

where $\text{E}^* \times \tilde{\text{E}}$ is the expectation with respect to $\text{P}^* \times \tilde{\text{P}}$.

The insurance company acts as a hedger of H in the financial market. It follows from (3.35) that the initial price of H is strictly less than that of the perfect hedge, since a survival probability is always less than one, or

$$_T U_x < \text{E}^* \left(S_T^2 + (S_T^1 - S_T^2)^+ \right) = \text{E}^* H.$$

Therefore, perfect hedging of H with an initial value of the hedge restricted by the fair price $\text{E}^* H$ is not possible and alternative hedging methods should be used. We will look for strategy π^* with some initial budget constraints such that its value $X_T^{\pi^*}$ at maturity is closed to H in the sense of the problem (3.2) with a power loss function (3.1). The strategy π^* is called the efficient hedge.

Once the efficient hedge is constructed, we will set the price of the equity-linked contract (3.33) as equal to its initial value $X_0^{\pi^*}$ and make conclusions about the appropriate balance between financial and insurance risk exposure.

Using Lemma 3.1, we reduce a construction of an efficient hedge for the claim H from (3.2) to an easier-to-do construction of a perfect hedge for the modified claim (3.4). We apply efficient hedging to equity-linked life insurance contracts.

Let us consider a single equity-linked life insurance with the payoff (3.33). From (3.34), we will pay attention to the term $(S_T^1 - S_T^2)^+ \cdot 1_{\{T(x)>T\}}$ associated with a call option. Note the following equality that comes from the definition of perfect and efficient hedging and Lemma 3.1:

$$X_0 = {_T p_x} \text{E}^*(S_T^1 - S_T^2)^+ = \text{E}^*(S_T^1 - S_T^2)_{p, p \geqslant 0}^+, \quad (3.36)$$

where $(S_T^1 - S_T^2)_p^+$ is defined by (3.4).

Using (3.36) we can separate insurance and financial components of the contract:

$$_T p_x = \frac{\text{E}^*(S_T^1 - S_T^2)_p^+}{\text{E}^*(S_T^1 - S_T^2)^+}. \quad (3.37)$$

The left-hand side of (3.37) is equal to the survival probability of the insured, which is a mortality risk for the insurer, while the right-hand side is related to a pure financial risk as it is connected to the evolution of the financial market. So, equation (3.37) as well as equations (2.57) and (3.7) can be viewed as the key balance equations combining the risk associated with the contract (3.33).

We use the efficient hedging methodology presented in Lemma 3.1 for a further development of the numerator of the right-hand side of (3.37) and the Margrabe formula for its denominator.

Let us first work with the denominator of the right-hand side of (3.37). We get

$$E^*(S_T^1 - S_T^2)^+ = S_0 \left\{ \Phi(b_+(1,1,T)) - \Phi(b_-(1,1,T)) \right\}, \qquad (3.38)$$

where $b_\pm(1,1,T) = \frac{\ln 1 \pm (\sigma_1 - \sigma_2)^2 \frac{T}{2}}{(\sigma_1 - \sigma_2)\sqrt{T}}$, $\Phi(x) = \frac{1}{\sqrt{2\pi}} \int_{-\infty}^{x} e^{-\frac{y^2}{2}} dy$.

The proof of (3.38) is given in the end of the section. Note that (3.38) is a version of the Margrabe formula for the case $S_0^1 = S_0^2 = S_0$, and $\rho = 1$.

To calculate the numerator of the right-hand size of (3.37), we want to represent it in terms of $Y_T = \frac{S_T^1}{S_T^2}$. Let us rewrite W_T with the help of a free parameter γ in the form

$$W_T = (1+\gamma)W_T - \gamma W_T$$

$$= \frac{1+\gamma}{\sigma_1}\left(\sigma_1 W_T + \left(\mu_1 - \frac{\sigma_1^2}{2}\right)T\right) - \frac{\gamma}{\sigma_2}\left(\sigma_2 W_T + \left(\mu_2 - \frac{\sigma_2^2}{2}\right)T\right) \qquad (3.39)$$

$$- \frac{1+\gamma}{\sigma_1}\left(\mu_1 - \frac{\sigma_1^2}{2}\right)T + \frac{\gamma}{\sigma_2}\left(\mu_2 - \frac{\sigma_2^2}{2}\right)T.$$

Using (3.30) and (3.39), we obtained the next representation of the density Z_T:

$$Z_T = G \cdot (S_T^1)^{\frac{(1+\gamma)\mu_1}{\sigma_1^2}} (S_T^2)^{\frac{\gamma\mu_1}{\sigma_1\sigma_2}}, \qquad (3.40)$$

where

$$G = (S_0^1)^{\frac{(1+\gamma)\mu_1}{\sigma_1^2}} (S_0^2)^{\frac{\gamma\mu_1}{\sigma_1\sigma_2}} \times$$

$$\times \exp\left\{ \frac{(1+\gamma)\mu_1}{\sigma_1^2}\left(\mu_1 - \frac{\sigma_1^2}{2}\right)T - \frac{\gamma\mu_1}{\sigma_1\sigma_2}\left(\mu_2 - \frac{\sigma_2^2}{2}\right)T - \frac{1}{2}\left(\frac{\mu_1}{\sigma_1}\right)^2 T \right\}.$$

Now we consider three cases according to (3.4) and choose appropriate values of the parameter γ for each case. The results are given in the following theorem.

Theorem 3.2. *Consider an insurance company measuring its shortfalls with a power loss function (3.1) with some parameter $p > 0$. For an equity-linked life insurance contract with the payoff (3.33) issued by the insurance company, it is possible to balance the survival probability of an insured and the financial risk associated with the contract.*

Case 1: $p > 1$
For $p > 1$ we get

$$TP_x = \frac{\Phi(b_+(1,C,T)) - \Phi(b_-(1,C,T))}{\Phi(b_+(1,1,T)) - \Phi(b_-(1,1,T))}$$
$$+ \frac{(C-1)^+}{C^{\alpha_p}} \exp\left\{\alpha_p(1-\alpha_p)\frac{(\sigma_1-\sigma_2)^2}{2}T\right\} \times \qquad (3.41)$$
$$\times \frac{\Phi(b_-(1,C,T) + \alpha_p(\sigma_1-\sigma_2)\sqrt{T})}{\Phi(b_+(1,1,T)) - \Phi(b_-(1,1,T))},$$

where C is found from

$$\alpha_p G^{\frac{1}{p-1}} C^{\alpha_p} = C - 1 \text{ and } \alpha_p = -\frac{\mu_1}{\sigma_1(\sigma_1-\sigma_2)(\rho-1)} - \frac{\sigma_2}{\sigma_1-\sigma_2}.$$

Case 2: $0 < p < 1$
Denote $\alpha_p = \frac{\sigma_1\sigma_2(1-\rho)-\mu_1}{\sigma_1(\sigma_1-\sigma_2)}$.
If $-\alpha_p \leq 1 - p$ (or $\frac{\mu_1}{\sigma_1^2} \leq 1 - \rho$) then

$$TP_x = 1 - \frac{\Phi(b_+(1,C,T)) - \Phi(b_-(1,C,T))}{\Phi(b_+(1,1,T)) - \Phi(b_-(1,1,T))}, \qquad (3.42)$$

where C is found from

$$C^{-\alpha_p} = a_p \cdot G \cdot \left((C-1)^+\right)^{1-p}. \qquad (3.43)$$

If $-\alpha_p > 1 - p$ (or $\frac{\mu_1}{\sigma_1^2} > 1 - \rho$) then

> If (3.43) has no solution then $TP_x = 1$.
>
> If (3.43) has one solution C, then TP_x is defined by (3.42).
>
> If (3.43) has two solutions $C_1 < C_2$ then

$$TP_x = 1 - \frac{\Phi(b_+(1,C_1,T)) - \Phi(b_-(1,C_1,T))}{\Phi(b_+(1,1,T)) - \Phi(b_-(1,1,T))}$$
$$+ \frac{\Phi(b_+(1,C_2,T)) - \Phi(b_-(1,C_2,T))}{\Phi(b_+(1,1,T)) - \Phi(b_-(1,1,T))}. \qquad (3.44)$$

Case 3: $p = 1$
For $p = 1$ we have

$$TP_x = 1 - \frac{\Phi(b_+(1,C,T)) - \Phi(b_-(1,C,T))}{\Phi(b_+(1,1,T)) - \Phi(b_-(1,1,T))}, \qquad (3.45)$$

where

$$C = (Ga_p)^{\frac{\sigma_1(\sigma_1-\sigma_2)}{\mu_1}} \text{ and } \alpha_p = -\frac{\mu_1}{\sigma_1(\sigma_1-\sigma_2)}.$$

The proofs of (3.41), (3.42), (3.44) and (3.45) are given after the following remark.

Remark 3.1. One can consider another approach to find C (or C_1 and C_2) for (3.42), (3.43) and (3.44). Let us fix the probability of the set $\{Y_T \leq C\}$ (or $\{Y_T \leq C_1\} \cup \{Y_T > C_2\}$):

$$P(Y_T \leq C) = 1 - \epsilon, \quad \epsilon > 0, \\ P(\{Y_T \leq C_1\} \cup \{Y_T > C_2\}) = 1 - \epsilon, \quad \epsilon > 0 \quad (3.46)$$

and calculate C (or C_1 and C_2) using log-normality of Y_T. Note that a set for which (3.46) is true coincides with $\{X_T^\pi \geq H\}$. The latter set has nice financial interpretation: fixing its probability at $1-\epsilon$, we specify the level of a financial risk that the company is ready to take or, in other words, the probability ϵ that it will not be able to hedge the claim perfectly.

Let us give the proof of the Margrabe formula (3.38) and Theorem 3.2.

Proof of (3.38). Denote $Y_T = \frac{S_T^1}{S_T^2}$ and find

$$\begin{aligned} E^*(S_T^1 - S_T^2)^+ &= E^*(S_T^1 - S_T^2) \cdot \mathbb{1}_{\{Y_T > 1\}} \\ &= E^* S_T^1 - E^* S_T^2 - E^*(S_T^1 - S_T^2) \cdot \mathbb{1}_{\{Y_T \leq 1\}}. \end{aligned} \quad (3.47)$$

Since S^1, S^2 are martingales with respect to P^*, we have $E^* S_T^i = S_0^i = S_0$, $i = 1, 2$. For the last term in (3.47), we get

$$E^* S_T^i \cdot \mathbb{1}_{\{Y_T \leq 1\}} = E^* \exp\{-\eta_i\} \cdot \mathbb{1}_{\{\xi \leq \ln 1\}}, \quad (3.48)$$

where $\eta_i = -\ln S_T^i$, $\xi = \ln Y_T$ are Gaussian random variables.

Using Lemma 2.3, we find that

$$E^* \exp\{-\eta_i\} \cdot \mathbb{1}_{\{\xi \leq \ln 1\}} = \exp\left\{\frac{\sigma_{\eta_i}^2}{2} - \mu_{\eta_i}\right\} \Phi\left(\frac{\ln 1 - (\mu_\xi - \text{cov}(\xi, \eta_i))}{\sigma_\xi}\right), \quad (3.49)$$

where $\mu_{\eta_i} = E^* \eta_i$, $\sigma_{\eta_i}^2 = \text{var}\,\eta_i$, $\mu_\xi = E^*\xi$, $\sigma_\xi^2 = \text{var}\,\xi$, $\Phi(x) = \frac{1}{\sqrt{2\pi}} \int\limits_{-\infty}^{x} e^{-\frac{y^2}{2}} dy$.

Using (3.48)–(3.49), we arrive to (3.38).

Proof of (3.41). According to (3.40), we have

$$Z_T^{\frac{1}{p-1}} (S_T^2)^{-1} = G^{\frac{1}{p-1}} \cdot (S_T^1)^{\frac{(1+\gamma)\mu_1}{\sigma_1^2(p-1)}} (S_T^2)^{-1 + \frac{\gamma \mu_1}{\sigma_1 \sigma_2(p-1)}} = G^{\frac{1}{p-1}} \cdot Y_T^{\alpha_p} \quad (3.50)$$

with

$$\alpha_p = -\frac{(1+\gamma)\mu_1}{\sigma_1^2(p-1)} = 1 - \frac{\gamma \mu_1}{\sigma_1 \sigma_2(p-1)}. \quad (3.51)$$

Equation (3.51) has the unique solution

$$\gamma = \gamma_p = \frac{\sigma_1^2 \sigma_2(p-1) + \mu_1 \sigma_2}{\mu_1(\sigma_1 - \sigma_2)}. \quad (3.52)$$

It follows from (3.52) that $\gamma_p > 0$ and, therefore, from (3.51) we conclude that $\alpha_p < 0$ and the equation

$$a_p G^{\frac{1}{p-1}} y^{\alpha_p} = (y-1)^+, \quad y \geqslant 1 \tag{3.53}$$

has the unique solution $C = C(p) \geqslant 1$. Using (3.50)–(3.53), we represent $(S_T^1 - S_T^2)_p^+$, as follows

$$(S_T^1 - S_T^2)_p^+ = S_T^2 (Y_T - 1)^+ - \left(a_p G^{\frac{1}{p-1}} Y_T^{\alpha_p} S_T^2\right) \wedge S_T^2 (Y_T - 1)^+$$
$$= S_T^2 \left((Y_T - 1)^+ - \left(a_p G^{\frac{1}{p-1}} Y_T^{\alpha_p}\right) \wedge (Y_T - 1)^+\right)$$
$$= S_T^2 \left((Y_T - 1)^+ - (Y_T - 1)^+ \mathbb{1}_{\{Y_T \leqslant C(p)\}} - a_p G^{\frac{1}{p-1}} Y_T^{\alpha_p} \mathbb{1}_{\{Y_T > C(p)\}}\right).$$

Taking into account that $\mathbb{1}_{\{Y_T > C(p)\}} = 1 - \mathbb{1}_{\{Y_T \leqslant C(p)\}}$, we get

$$\begin{aligned} \mathrm{E}^*(S_T^1 - S_T^2)_p^+ &= \mathrm{E}^*(S_T^1 - S_T^2)^+ - \mathrm{E}^*(S_T^1 - S_T^2)^+ \mathbb{1}_{\{Y_T \leqslant C(p)\}} \\ &\quad - a_p G^{\frac{1}{p-1}} \left(\mathrm{E}^* S_T^2 Y_T^{\alpha_p} - \mathrm{E}^* S_T^2 Y_T^{\alpha_p} \mathbb{1}_{\{Y_T \leqslant C(p)\}}\right). \end{aligned} \tag{3.54}$$

Since $C(p) \geqslant 1$, we have

$$\begin{aligned} \mathrm{E}^*(S_T^1 - S_T^2)^+ - \mathrm{E}^*(S_T^1 - S_T^2)^+ \mathbb{1}_{\{Y_T \leqslant C(p)\}} &= \mathrm{E}^*(S_T^1 - S_T^2) \mathbb{1}_{\{Y_T > C(p)\}} \\ &= \mathrm{E}^*(S_T^1 - S_T^2) - \mathrm{E}^*(S_T^1 - S_T^2) \mathbb{1}_{\{Y_T \leqslant C(p)\}}. \end{aligned} \tag{3.55}$$

Using (3.55), we can calculate the difference between the first two terms in (3.54) reproducing exactly the same procedure as in (3.47)–(3.49) and arrive at the equality

$$\begin{aligned} \mathrm{E}^*(S_T^1 - S_T^2)^+ &- \mathrm{E}^*(S_T^1 - S_T^2)^+ \mathbb{1}_{\{Y_T \leqslant C(p)\}} \\ &= S_0 \{\Phi(b_+(1, C, T)) - \Phi(b_-(1, C, T))\}. \end{aligned} \tag{3.56}$$

To calculate the other two terms in (3.54), we represent the product $S_T^2 Y_T^{\alpha_p}$ as follows

$$\begin{aligned} S_T^2 Y_T^{\alpha_p} &= S_0 \times \exp\left\{(\sigma_1 \alpha_p + \sigma_2(1 - \alpha_p)) W_T^* - \frac{1}{2}\left(\sigma_1^2 \alpha_p + \sigma_2^2(1 - \alpha_p)\right) T\right\} \\ &= S_0 \times \exp\left\{(\sigma_1 \alpha_p + \sigma_2(1 - \alpha_p)) W_T^* - \frac{1}{2}(\sigma_1 \alpha_p + \sigma_2(1 - \alpha_p))^2 T \right.\\ &\quad \left. + \frac{1}{2}(\sigma_1 \alpha_p + \sigma_2(1 - \alpha_p))^2 T - \frac{1}{2}\left(\sigma_1^2 \alpha_p + \sigma_2(1 - \alpha_p)\right) T\right\} \\ &= S_0 \times \exp\left\{(\sigma_1 \alpha_p + \sigma_2(1 - \alpha_p)) W_T^* - \frac{1}{2}(\sigma_1 \alpha_p + \sigma_2(1 - \alpha_p))^2 T \right. \\ &\quad \left. - \alpha_p(1 - \alpha_p)(\sigma_1 - \sigma_2)^2 \frac{T}{2}\right\}. \end{aligned} \tag{3.57}$$

Taking an expected value of (3.57) with respect to P*, we find that

$$E^* S_T^2 Y_T^{\alpha_p} = S_0 \exp\left\{-\alpha_p(1-\alpha_p)(\sigma_1-\sigma_2)^2 \frac{T}{2}\right\}. \tag{3.58}$$

Using (3.57)–(3.58) and following the same steps as in (3.47)–(3.49), we obtain

$$-\alpha_p G^{\frac{1}{p-1}} \left(E^* S_T^2 Y_T^{\alpha_p} - E^* S_T^2 Y_T^{\alpha_p} \mathbb{1}_{\{Y_T \leqslant C(p)\}}\right)$$
$$= -\alpha_p G^{\frac{1}{p-1}} S_0 \exp\left\{-\alpha_p(1-\alpha_p)(\sigma_1-\sigma_2)^2 \frac{T}{2}\right\} \times \tag{3.59}$$
$$\times \Phi(b_-(1,C,T) + \alpha_p(\sigma_1-\sigma_2)\sqrt{T}).$$

Combining (3.37), (3.38), (3.54), (3.56), and (3.59), we arrive (3.41).

Proof of (3.42) *and* (3.44). Taking into account the structure of $(S_T^1-S_T^2)_p^+$ in (3.4), we represent the product $Z_T(S_T^2)^{1-p}$ with the help of a free parameter γ (see (3.39), (3.40), (3.50)–(3.53)) and get

$$Z_T(S_T^2)^{1-p} = G(S_T^1)^{\frac{(1+\gamma)\mu_1}{\sigma_1^2}} (S_T^2)^{\frac{\gamma\mu_1}{\sigma_1\sigma_2}} (S_T^2)^{1-p} = G Y_T^{\alpha_p}, \tag{3.60}$$

where

$$\alpha_p = -\frac{(1+\gamma)\mu_1}{\sigma_1^2} = -\frac{\gamma\mu_1}{\sigma_1\sigma_2} - (1-p)$$

and hence,

$$\gamma = \gamma_p = \frac{\sigma_2(\mu_1 - (1-p)\sigma_1^2)}{\mu_1(\sigma_1-\sigma_2)},$$
$$-\alpha_p = \frac{\mu_1}{\sigma_1^2}\left(1 + \frac{\sigma_2(\mu_1-(1-p)\sigma_1^2)}{\mu_1(\sigma_1-\sigma_2)}\right) \tag{3.61}$$
$$= \frac{\mu_1}{\sigma_1^2} + \frac{\sigma_2}{\sigma_1-\sigma_2}\left(\frac{\mu_1}{\sigma_1^2} - (1-p)\right).$$

Consider the following characteristic equation:

$$\gamma^{-\alpha_p} \alpha_p G((y-1)^+)^{1-p}, \quad y \geqslant 0. \tag{3.62}$$

If $-\alpha_p > 1-p$, then according to (3.61)

$$\frac{\mu_1}{\sigma_1^2} + \frac{\sigma_2}{\sigma_1-\sigma_2}\left(\frac{\mu_1}{\sigma_1^2} - (1-p)\right) > 1-p \text{ or } \frac{\mu_1}{\sigma_1^2} > 1-p > 0. \tag{3.63}$$

In this case equation (3.62) has zero, one, or two solutions. All these situations can be considered in way similar to Case 1.

If (3.62) has no solution then $\mathbb{1}_{\{Z_T^{-1} > \alpha_p H^{1-p}\}} \equiv 1$, $H_p = H$ and therefore, $Tp_x = 1$.

If equation (3.62) has only one solution $C = C(p)$ then $(S_T^1 - S_T^2)_p^+ = (S_T^1 - S_T^2)^+ \mathbb{1}_{\{Y_T \leqslant C(p)\}}$ and, according to (3.37), we arrive at (3.41).

If there are two solutions $C_1(p) < C_2(p)$ to (3.62) then the structure of the modified claim is $(S_T^1 - S_T^2)_p^+ = (S_T^1 - S_T^2)^+ \mathbb{1}_{\{Y_T \leq C_1(p)\}} + (S_T^1 - S_T^2)^+ \mathbb{1}_{\{Y_T > C_2(p)\}}$ and we arrive at (3.44).

If $-\alpha_p \leq 1 - p$, then $\frac{\mu_1}{\sigma_1^2} \leq 1 - p < 1$ and therefore, equation (3.62) has only one solution $C = C(p)$. This is equivalent to a previous case with one solution and reproducing the same reasons, we arrive at (3.42).

Proof of (3.45). According to (3.40), we represent the density Z_T as follows

$$Z_T = G(S_T^1)^{\frac{(1+\gamma)\mu_1}{\sigma_1^2}} (S_T^2)^{\frac{\gamma\mu_1}{\sigma_1\sigma_2}} = GY_T^{\alpha_p}, \qquad (3.64)$$

where

$$\alpha_p = -\frac{(1+\gamma)\mu_1}{\sigma_1^2} = -\frac{\gamma\mu_1}{\sigma_1\sigma_2}$$

and, therefore,

$$\gamma = \gamma_p = \frac{\sigma_1\sigma_2}{\sigma_1(\sigma_1 - \sigma_2)},$$

$$-\alpha_p = \frac{\mu_1}{\sigma_1\sigma_2}\gamma_p = \frac{\mu_1}{\sigma_1(\sigma_1 - \sigma_2)}, \quad \sigma_1 > \sigma_2.$$

From (3.40) and (3.64) we find

$$(S_T^1 - S_T^2)^+ \mathbb{1}_{\{Y_T^{-\alpha_p} > G\alpha_p\}} = (S_T^1 - S_T^2)^+ \mathbb{1}_{\left\{Y_T^{\frac{\mu_1}{\sigma_1(\sigma_1-\sigma_2)}} > G a_p\right\}} \qquad (3.65)$$

$$= (S_T^1 - S_T^2)^+ \mathbb{1}_{\{Y_T > C\}},$$

where

$$C = (G a_p)^{\frac{\sigma_1(\sigma_1-\sigma_2)}{\mu_1}}.$$

Using (3.37), (3.38), (3.55), (3.56), and (3.65), we arrive at (3.45).

3.3 Illustrative risk-management for the risk-taking insurer

Here we will explore (3.46) and Remark 3.1 to demonstrate how efficient hedging works for risk-management needs. The loss function with $p > 1$ corresponds to a company avoiding risk with risk aversion increasing as p grows. The case $0 < p < 1$ is appropriate for companies that are inclined to take some risk. We show how a risk-taking insurance company could use efficient hedging for management of its financial and insurance risk. For illustrative purposes we consider the extreme case when $p \to 0$. While the effect of a power p close to zero on efficient hedging was pointed out before, we give a different interpretation and implementation which are better suited for the purposes of our analysis. In addition, we restrict our attention to a particular case for which equation (3.42) has only one solution. This is done for illustrative purposes only since the calculation of constants C, C_1 and C_2 for other cases may involve the use of extensive numerical techniques and lead us well beyond our research purposes.

As was mentioned above, the characteristic equation (3.42) with $p \leqslant 1+\alpha_p$ (or equivalently, $p \leqslant 1 - \frac{\mu_1}{\sigma_1^2}$) admits only one solution C which is further used for determination of a modified claim (3.4) as follows

$$H_p = H \cdot \mathbb{1}_{\{Y_T \leqslant C\}},$$

where $H = (S_T^1 - S_T^2)^+$, $Y_T = \frac{S_T^1}{S_T^2}$, and $0 < p < 1$.

Denote an efficient hedge for H and its initial value as π^* and $x = X_0$, respectively. It follows from Lemma 3.1 that π^* is a perfect hedge for $H_p = (S_T^1 - S_T^2)_p^+$.

Since the inequality $((a-b)^+)^p \leqslant a^p$ is true for any positive a and b, we have

$$\mathrm{E}\left((H - X_T^{\pi^*}(x))^+\right)^p$$
$$= \mathrm{E}\left[(H_p - X_T^{\pi^*}(x))^+ \cdot \mathbb{1}_{\{Y_T \leqslant C\}} + (H - X_T^{\pi^*}(x))^+ \cdot \mathbb{1}_{\{Y_T > C\}}\right]^p$$
$$= \mathrm{E}\left[(H - X_T^{\pi^*}(x))^+ \cdot \mathbb{1}_{\{Y_T > C\}}\right]^p \qquad (3.66)$$
$$= \mathrm{E}\left[(H - X_T^{\pi^*}(x))^+\right]^p \cdot \mathbb{1}_{\{Y_T > C\}} \leqslant \mathrm{E} H^p \cdot \mathbb{1}_{\{Y_T > C\}}.$$

Taking the limit in (3.66) as $p \to 0$ and applying the classical Lebesgue dominated convergence theorem, we obtain

$$\mathrm{E} H^p \cdot \mathbb{1}_{\{Y_T > C\}} \xrightarrow{p \to 0} \mathrm{E} \mathbb{1}_{\{Y_T > C\}} = \mathrm{P}(Y_T > C). \qquad (3.67)$$

Therefore, we can fix a probability $\mathrm{P}(Y_T > C) = \epsilon$ which quantifies a financial risk and is equivalent to the probability of failing to hedge H at maturity.

Note that the same hedge π^* will also be an efficient hedge for the claim $b \cdot H$, where b is some positive constant but its initial value will be $b \cdot x$ instead of x. We will use this simple observation for pricing cumulative claims below when we consider the insurance company taking advantage of diversification of a mortality risk and further reducing the price of the contract.

Here, we pool together the homogeneous clients of the same age, life expectancy and investment preferences and consider a cumulative claim $l_{x+T} \cdot H$, where l_{x+T} is the number of insured alive at time T from the group of size l_x. Let us measure as in Section 2.3 the mortality risk of a pool of the equity-linked life insurance contracts for this group with the help of a parameter $\delta \in (0, 1)$ such that

$$\widetilde{\mathrm{P}}(l_{x+T} \leqslant n_\delta) = 1 - \delta, \qquad (3.68)$$

where n_δ is some constant.

In other words, δ equals the probability that the number of clients alive at maturity will be greater than expected based on the life expectancy of homogeneous clients. Since it follows a frequency distribution, this probability could be calculated with the help of a binomial distribution with parameters $_Tp_x$ and l_x where $_Tp_x$ is found by fixing the level of the financial risk ϵ and applying the formulate from Theorem 3.2.

We can rewrite (3.68) as follows

$$\tilde{P}\left(\frac{l_{x+T}}{l_x} \leqslant \frac{n_\delta}{l_x}\right) = \tilde{P}\left(\frac{l_{x+T}}{l_x} \leqslant b\right) = 1 - \delta,$$

where $b = \frac{n_\delta}{l_x}$.

Due to the independence of insurance and financial risk, we have

$$P \times \tilde{P}\left(l_x X_T^\pi(bx) \geqslant l_{x+T} H\right) \geqslant P \times \tilde{P}\left(X_T^\pi(bx) \geqslant \frac{l_{x+T}}{l_x} H\right)$$

$$\geqslant P\left(X_T^\pi(bx) \geqslant bH\right) \cdot \tilde{P}\left(n_\delta \geqslant l_{x+T}\right) \geqslant (1 - \epsilon)(1 - \delta) \geqslant 1 - (\epsilon + \delta).$$

So, using strategy π^* the insurance company is able to hedge the cumulative claim $l_{x+T} \cdot H$ with the probability at least $1 - (\epsilon + \delta)$ which combines both financial and insurance risks. The price of a single contract will be further reduced to $\frac{n_\delta}{l_x} {}_Tp_x E^* H$.

Example 3.2. Using the same reasons as before, we restrict our attention to the case when $p \to 0$ and the equation (3.43) has only one solution. Consider the following parameters for the risky assets:

$$\mu_1 = 5\%, \quad \sigma_1 = 23\%,$$
$$\mu_2 = 4\%, \quad \sigma_2 = 19\%.$$

The condition (3.31) is approximately fulfilled to preclude the existence of arbitrage opportunities. Also, since $1 - \frac{\mu_1}{\sigma_1} \cong 0.05$, p should be very small, or $p \leqslant 0.05$, and we are able to use (3.67) and exploit (3.42) from Theorem 3.2. For survival probabilities we use the Uninsured Pensioner Mortality Table UP-94 which is based on best estimate assumptions for mortality. Further, we assume that a unit of equity-linked life insurance contracts has the initial value $S_0 = 100$. We consider contracts with maturity terms $T = 5, 10, 15, 20, 25$ years. The number of homogeneous insured in a cohort is $l_x = 100$.

Figure 3.1 represents the offsetting relationships between financial and insurance risk. Note that financial and insurance risks do offset others. As perfect hedging is impossible, the insurer will be exposed to a financial risk expressed as a probability that it will be unable to hedge the claim (3.34) with the probability one. At the same time, the insurance company faces a mortality risk or probability that the insured will be alive at maturity and the payment (3.34) will be due at that time. Combining both risks together, we conclude that if the financial risk is big, the insurance company may prefer

Valuation of equity-linked life insurance contracts

to be exposed to a smaller mortality risk. By contracts, if the claim (3.34) could be hedged with greater probability, the insurance company may wish to increase its mortality risk exposure. Therefore, there is an offset between financial and mortality risks the insurer can play with: by fixing one of the risks, the appropriate level of another risk could be calculated.

In figure 3.1 we obtained survival probabilities using (3.42) for different levels of a financial risk ϵ and found the corresponding ages for the clientele using the specified mortality table. Note that wherever the risk that the insurance company will fail to hedge successfully increases, the recommended ages for clients rise as well. As a result, the company diminishes the insurance component of risk by attracting older and, therefore, safer clientele to compensate for the increasing financial risk. Also, observe that with longer contract maturities, the company can widen its audience to younger clients because a mortality risk, which is a survival probability in our case, is decreasing over time.

Different combinations of a financial risk ϵ and an insurance risk δ give us the range of price for the equity-linked contracts. The results for the contracts are shown in Figure 3.2.

The next step is to construct a grid that enables the insurance company to identify the acceptable level of the financial risk for insured of any age. We restrict our attention to a group of clients of ages 30, 40, 50, and 60 years. The results are presented in Table 3.4. The financial risk found reflects the probability of failure to hedge the payoff that will be offset by the mortality risk of the clients of a certain age. Prices of the contracts for the same group of clients are given in Table 3.5. Note that the price of a contract is a function of financial and insurance risks associated with this contract. The level of the insurance risk is chosen to be $\delta = 2.5\%$. In the last row, the Margrabe prices are compared with the reduced prices of equity-linked contracts. The reduction in prices was possible for two reasons: we took into account the mortality risk of an individual client (the probability that the client would not survive to maturity and, therefore, no payment at maturity would be made) and the possibility to diversify the cumulative mortality risk by pooling homogeneous clients together.

Age of Clients	$T = 5$	$T = 10$	$T = 15$	$T = 20$	$T = 25$
30	0.05%	0.13%	0.25%	0.45%	0.8%
40	0.1%	0.25%	0.55%	1.2%	2.3%
50	0.2%	0.7%	1.8%	3.7%	7%
60	0.8%	2.5%	5.5%	10.5%	18.5%

TABLE 3.4: Acceptable financial risk offsetting mortality risk of individual clients.

Age of Clients	$T=5$	$T=10$	$T=15$	$T=20$	$T=25$
30	3.45	4.86	5.87	6.66	7.22
40	3.45	4.79	5.69	6.25	6.45
50	3.39	4.56	5.11	5.10	4.53
60	3.17	3.84	3.76	2.99	1.70
Margrabe price	3.57	5.04	6.17	7.13	7.97

TABLE 3.5: Prices of contracts with cumulative mortality risk $\delta = 2.5\%$.

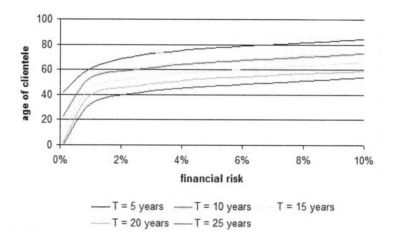

FIGURE 3.1: Offsetting financial and mortality risks.

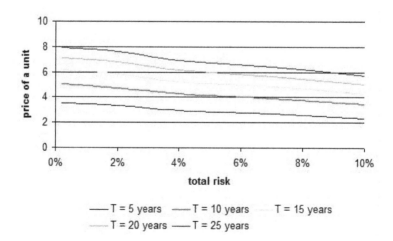

FIGURE 3.2: Prices of $100 invested in equity-linked life insurance contracts.

Chapter 4

Quantile hedging and risk management of contracts for diffusion and jump-diffusion models

4.1 Quantile pricing formulas for contracts with perfectly correlated stocks and guarantees: Diffusion and jump-diffusion settings ... 77

4.2 Quantile hedging and an exemplary risk-management scheme for equity-linked life insurance 90

4.1 Quantile pricing formulas for contracts with perfectly correlated stocks and guarantees: Diffusion and jump-diffusion settings

We develop here the method for quantile hedging in the framework of two-factor diffusion and jump-diffusion market models. We start from the Black–Scholes model studied in Paragraph 3.2 in efficient hedging aspects. So, we consider the model (see (3.29)):

$$dS_t^i = S_t^i(\mu_i dt + \sigma_i dW_t), \quad i = 1, 2, t \leq T. \quad (4.1)$$

To simplify calculations we assume that interest rate $r = 0$. An insurance company that operates on the financial market issues a contract linked to the evolution of risky assets S^1 and S^2, where S^1 is supposed to be more risky, or $\sigma_1 > \sigma_2$. Let us define H as $\max\{S_T^1, S_T^2\}$. Therefore, S^2 plays the role of a flexible guarantee.

As in Chapters 2 and 3 we use a random variable $T(x)$ on some probability space $(\widetilde{\Omega}, \widetilde{\mathcal{F}}, \widetilde{P})$ to represent the remaining lifetime of an insured of age x. We consider a pure endowment contract with the payoff function

$$H(T(x)) = H \cdot \mathbb{1}_{\{T(x) > T\}}. \quad (4.2)$$

To find the initial price of the contract (4.2), we take its expected value with respect to $P^* \times \widetilde{P}$ and arrive at the Brennan–Schwartz price:

$$\begin{aligned}{}_TU_x &= E^* \times \widetilde{E}\left[H(T(x))\right] = E^*[H] \cdot \widetilde{E}\mathbb{1}_{\{T(x) > T\}} \\ &= E^*[H] \cdot {}_Tp_x = S_0^2 {}_Tp_x + E^*\left(S_T^1 - S_T^2\right)^+ \cdot {}_Tp_x.\end{aligned} \quad (4.3)$$

The difference $_TU_x - S_0^2{}_TP_x$ from (4.3) can be viewed as a bound for the initial value of a hedging strategy for the embedded option $(S_T^1 - S_T^2)^+$. We have the following analog of the balance equation (2.57) and (3.37) when we apply quantile hedging to (4.2) with the embedded option instead

$$_TP_x = \frac{\mathrm{E}^* \left(S_T^1 - S_T^2\right)^+ \cdot \mathbb{1}_{A^*}}{\mathrm{E}^* \left(S_T^1 - S_T^2\right)^+}, \qquad (4.4)$$

where A^* is the maximal set of successful hedging for $(S_T^1 - S_T^2)^+$.

Recall (see (3.32)) that in the model (4.1) the second risky asset, S^2, can be expected through S^1:

$$S_T^2 = (S_0^1)^{-\frac{\sigma_2}{\sigma_1}} S_0^2 \exp\left\{\left[\mu_2 - \frac{\sigma_2}{\sigma_1}\mu_1 + \frac{\sigma_2}{2}(\sigma_1 - \sigma_2)\right]T\right\} (S_T^1)^{\frac{\sigma_2}{\sigma_1}}. \qquad (4.5)$$

To make S^1 and S^2 martingales with respect to P*, we assume that the following condition is satisfied (see (3.31)):

$$\sigma_2 \mu_1 - \sigma_1 \mu_2 = 0. \qquad (4.6)$$

Such a measure P*, under condition (4.6) is given by density

$$Z_T = \exp\left\{-\frac{\mu_1}{\sigma_1} W_T - \frac{1}{2}\left(\frac{\mu_1}{\sigma_1}\right)^2 T\right\}. \qquad (4.7)$$

We reduce pricing of (4.2) to the pricing of another contract

$$(S_T^1 - S_T^2)^+ \cdot \mathbb{1}_{\{T(x) > T\}}$$

and reproduce an analysis similar to most of Paragraph 3.2, using another representation of W_T and using a free parameter γ (3.39):

$$W_T = (1+\gamma)W_T - \gamma W_T$$
$$= \frac{1+\gamma}{\sigma_1}\left[\sigma_1 W_T + \left(\mu_1 - \frac{\sigma_1^2}{2}\right)T\right] - \frac{\gamma}{\sigma_2}\left[\sigma_2 W_T + \left(\mu_2 - \frac{\sigma_2^2}{2}\right)T\right] \qquad (4.8)$$
$$- \frac{1+\gamma}{\sigma_1}\left[\mu_1 - \frac{\sigma_1^2}{2}T\right] + \frac{\gamma}{\sigma_2}\left[\mu_2 - \frac{\sigma_2^2}{2}\right]T.$$

According to the Neyman–Pearson lemma, the set A^* has the following structure:

$$A^* = \{Z_T^{-1} \geq a(S_T^1 - S_T^2)^+\} = \left\{(Z_T S^2)^{-1} \geq a\left(\frac{S_T^1}{S_T^2} - 1\right)^+\right\}, \qquad (4.9)$$

where a is a constant.

Using (4.5), (4.7)–(4.8), we represent $Z_T S_T^2$ in (4.9) as follows:

$$Z_T S_T^2 = Y_T^\alpha G,$$

where

$$G = (S_0^1)^{\frac{(1+\gamma)\mu_1}{(\sigma_1)^2}} (S_0^2)^{\frac{\gamma\mu_1}{\sigma_1 \sigma_2}} \times$$

$$\times \exp\left\{-\frac{1+\gamma}{\sigma_1}\left[\mu_1 - \frac{1}{2}(\sigma_1)^2\right]T - \frac{\gamma\mu_1}{\sigma_1\sigma_2}\left[\mu_2 - \frac{1}{2}(\sigma_2)^2\right]T - \frac{1}{2}\left(\frac{\mu_1}{\sigma_1}\right)^2 T\right\},$$

$$\alpha = -\frac{(1+\gamma)\mu_1}{(\sigma_1)^2} = -\frac{\gamma\mu_1}{\sigma_1\sigma_2} - 1. \tag{4.10}$$

Assuming $0 < \sigma_1 - \sigma_2 \ll \sigma_1$ and σ_2, we choose $\gamma = \sigma_2 - \sigma_1$ and obtain the following sufficient condition for (4.10):

$$\mu_1 = \frac{\sigma_1^2 \sigma_2}{\sigma_2 + (\sigma_1 - \sigma_2)^2}. \tag{4.11}$$

Now we consider the characteristic equation:

$$x^{-\alpha} = G \cdot \alpha \cdot (x-1)^+. \tag{4.12}$$

Due to (4.11), we obtain

$$-\alpha = \frac{\sigma_2 + (\sigma_2 - \sigma_1)\sigma_2}{\sigma_2 - (\sigma_1 - \sigma_2)^2} < 1,$$

and therefore, equation (4.12) admits the only solution c.

Now we will formulate the theorem representing the main result of the application of quantile hedging methodology to equity-linked life insurance contracts.

Theorem 4.1. *Consider a financial market with two risky assets following the model (4.1) for which (4.6) and (4.11) are true and an equity-linked life insurance contract with the payoff $(S_T^1 - S_T^2)^+ \cdot \mathbb{1}_{\{T(x)>T\}}$. Then the following is true for a survival probability of an insured:*

$$_T p_x = 1 - \frac{S_0^1 \Phi(b_+(S_0^1, cS_0^2, T)) - S_0^2 \Phi(b_-(S_0^1, cS_0^2, T))}{S_0^1 \Phi(b_+(S_0^1, S_0^2, T)) - S_0^2 \Phi(b_-(S_0^1, S_0^2, T))}. \tag{4.13}$$

Proof. Consider the set $A^* = \{Y_T \leq c\}$. Rewriting (4.1) in the exponential form of geometric Brownian motion, we can represent Y_T as follows

$$Y_T = \frac{S_0^1}{S_0^2} \exp\left\{(\sigma_1 - \sigma_2)W_T + \left[(\mu_1 - \mu_2) - \frac{\sigma_1^2 - \sigma_2^2}{2}\right]T\right\}$$

$$= \frac{S_0^1}{S_0^2} \exp\left\{(\sigma_1 - \sigma_2)W_T^* - \frac{(\sigma_1^2 - \sigma_2^2)T}{2}\right\},$$

where $W_t^* = W_t + \frac{\mu_1}{\sigma_1}t$ is a new Wiener process with respect to P*.

It follows from the above formula for Y_T that $\ln(Y_T)$ is normally distributed:

$$N\left(\ln\frac{S_0^1}{S_0^2} - \frac{(\sigma_1^2 - \sigma_2^2)T}{2}, (\sigma_1 - \sigma_2)^2 T\right) = N(\mu_{\ln Y_T}, \sigma_{\ln Y_T}^2)$$

with respect to P*, and

$$N\left(\ln\frac{S_0^1}{S_0^2} + \left[\mu_1 - \mu_2 - \frac{\sigma_1^2 - \sigma_2^2}{2}\right]T, (\sigma_1 - \sigma_2)^2 T\right)$$

with respect to P.

Therefore, fixing $P(A^*) = 1 - \epsilon$, $\epsilon > 0$, we may have the following equation for c through normal distribution with some parameters μ and σ^2:

$$1 - \epsilon = P(A^*) = P\{\ln Y_T \leq \ln c\} = \Phi_{\mu,\sigma^2}(\ln c). \tag{4.14}$$

The denominator of (4.4) is given by the Margrabe formula

$$\mathbb{C}^{\text{Mar}}(S_0^1, S_0^2, T) = S_0^1 \Phi(b_+(S_0^1, S_0^2, T)) - S_0^2 \Phi(b_-(S_0^1, S_0^2, T)). \tag{4.15}$$

For the numerator (4.4) we obtain

$$\begin{aligned}E^*\left[(S_T^1 - S_T^2)^+ \mathbb{1}_{A^*}\right] &= E^*\left[(S_T^1 - S_T^2)^+ \mathbb{1}_{\{Y_T \leq c\}}\right] \\ E^*\left[(S_T^1 - S_T^2)^+\right] &- (S_0^1 - S_0^2) + E^*\left[(S_T^1 - S_T^2)\mathbb{1}_{\{Y_T \leq c\}}\right].\end{aligned} \tag{4.16}$$

To determine $E^* S_T^i \mathbb{1}_{\{Y_T \leq c\}}$, $i = 1, 2$, in (4.16), we use Lemma 2.3 which leads us together with (4.15) to (4.13). □

Example 4.1. Consider the financial indices of the Russell 2000 (RUT-I) and the Dow Jones Industrial Average (DJIA) as risky assets S^1 and S^2. The Russell 2000 is the index of small US companies' stocks, whereas the Dow Jones Industrial Average is based on the portfolio consisting of 30 blue-chip stocks in the United States. The first index, RUT-I, is supposed to be more risky than DJIA. We estimate (μ_1, σ_1) and (μ_2, σ_2) the rates of returns and volatilities for RUT-I and DJIA empirically, using daily observations of prices from August 1, 1997, to July 31, 2003. We obtain the following numbers:

$$\mu_1 = 0.0481, \quad \sigma_1 = 0.2232;$$
$$\mu_2 = 0.0417, \quad \sigma_2 = 0.2089.$$

We observe that the conditions (4.6) and (4.11) with $\gamma = \sigma_2 - \sigma_1$ are approximately fulfilled. We make the initial values of both assets equal. Utilizing the formulae (4.13) and (4.14) with $T = 1, 3, 5, 10$ and $\epsilon = 0.01, 0.025, 0.05$, we find the values of the corresponding survival probabilities $_Tp_x$ (see Table 4.1). Now we can find an age of the insured in life tables, (see, for instance, Bowers et al.(1997)). The data is displayed in Table 4.2.

Quantile hedging and risk management of contracts

	$\epsilon = 0.01$	$\epsilon = 0.025$	$\epsilon = 0.05$
$T = 1$	0.9447	0.8774	0.7811
$T = 3$	0.9511	0.8910	0.8041
$T = 5$	0.9549	0.8989	0.8174
$T = 10$	0.9605	0.9108	0.8378

TABLE 4.1: Survival probabilities.

	$\epsilon = 0.01$	$\epsilon = 0.025$	$\epsilon = 0.05$
$T = 1$	78	87	94
$T = 3$	61	71	79
$T = 5$	53	63	71
$T = 10$	41	50	58

TABLE 4.2: Age of insured.

When the level of risk ϵ increases, the company should restrict the group of insured by attracting older clients. As a result, the company diminishes the insurance component of risk to compensate for increasing financing risk. Issuing contracts for a longer term T allows the insurance company to diminish insurance risk with fixed ϵ. Therefore, the company can afford to work with younger groups of clients.

In additional, the insurance company can diversify a mortality risk by pooling homogeneous clients together. We consider the cumulative claim $l_{x+T}(S_T^1 - S_T^2)^+$, where l_{x+T} is the number of insured at the end of the contract from the group of size l_x. Denote by $\pi = \pi_\epsilon$ a quantile hedge of the risk level ϵ with initial, (quantile), price C_ϵ and terminal value X_T^π so that

$$P\left(X_T^\pi \geqslant (S_T^1 - S_T^2)^+\right) = 1 - \epsilon. \tag{4.17}$$

The maximal set of successful hedging is invariant with respect to multiplication by a positive constant b. Hence, the claim $b(S_T^1 - S_T^2)^+$ can be hedged at the same risk level ϵ with initial price bC_ϵ. Let's provide here the same arguments to combine both financial and insurance risks together as in Paragraph 2.3 (see (2.88)–(2.90)) and in Paragraph 3.3 (see (3.68) and related formulas). Take $b = \frac{n_\delta}{l_x}$, where the number n_δ is determined from the equality

$$P(l_{x+T} \leqslant n_\delta) = 1 - \delta. \tag{4.18}$$

The parameter $\delta \in (0,1)$ characterizes the level of mortality risk of the company, and the probability in (4.18) can be computed with the help of the binomial distribution with parameter $_T p_x$. Using the independence of l_{x+T}

and the market assets S^1 and S^2 we derive from (4.17) and (4.18) that

$$P \times \tilde{P}\left(l_x X_T^\pi \geqslant l_{x+T}(S_T^1 - S_T^2)^+\right) \geqslant P \times \tilde{P}\left(X_T^\pi \geqslant \frac{l_{x+T}}{l_x}(S_T^1 - S_T^2)^+\right)$$

$$\geqslant P\left(X_T^\pi \geqslant \frac{n_\delta}{l_x}(S_T^1 - S_T^2)^+\right) \cdot \tilde{P}\left(l_{x+T} \leqslant n_\delta\right)$$

$$\geqslant (1-\epsilon)(1-\delta) \geqslant 1 - (\epsilon + \delta). \tag{4.19}$$

Let us take $T = 1, 5, 10$, fix both risk levels $\epsilon = \delta = 0.025$, and consider the contract for the group of size $l_x = 100$. Due to (4.19), diversification of a mortality risk allows the insurance company to reduce the price of a single contract by 12% to 18% in comparison with the Margrabe prices while the corresponding quantile price is reduced by 9% to 12%.

Let us extend the model (4.1) adding a jump component in the evolution of stock prices. We consider a financial market with two risky assets, S^1 and S^2, described by a two-factor jump-diffusion model:

$$dS_t^i = S_{t-}^i(\mu_i dt + \sigma_i dW_t - \nu_i d\Pi_t), \quad i = 1, 2, \tag{4.20}$$

where $\mu_i \in \mathbb{R}$, $\sigma_i > 0$, $\nu_i < 1$.

We suppose that all processes are defined on a standard stochastic basis $(\Omega, \mathcal{F}, \mathbb{F} = (\mathcal{F}_t)_{t \geqslant 0}, P)$ and are adapted to the filtration $\mathbb{F}$, generated by a Wiener process W and Poisson process Π with intensity $\lambda > 0$, whose paths are right-continuous with finite left limits at each $t > 0$ (see Chapter 1). We assume that S^1 is riskier than S^2, or $\sigma_1 > \sigma_2$.

We consider pure endowment life insurance contracts linked to the evolution of both risky assets. The payoff at maturity is $\max\{S_T^1, S_T^2\}$, where T is the maturity time, provided that an insured is alive at maturity. As a result, the contract under our consideration has a flexible guarantee S^2 and a potential for future gains associated with S^1.

The risky asset S^i is defined by its price process S_t^i, $t \geqslant 0$, $i = 1, 2$. For the sake of simplicity, we assume that the non-risky asset $B_t \equiv 1$. Hence, the interest rate $r = 0$.

A predictable process $\pi = (\pi_t)_{t \geqslant 0} = (\beta_t, \gamma_t^1, \gamma_t^2)_{t \geqslant 0}$ is called a trading strategy, or a portfolio. The value (capital) of π at time is equal to

$$X_t^\pi = \beta_t + \sum_{i=1}^{2} \gamma_t^i S_t^i.$$

The class of portfolios π whose value evolves as

$$X_t^\pi = X_0^\pi + \sum_{i=1}^{2} \int_0^t \gamma_u^i S_u^i$$

is denoted SF. We call π self-financing if $\pi \in SF$. We consider admissible portfolios $\pi \in SF$ whose capital is positive.

The market (4.20) admits a single martingale measure P* if (see details in Chapter 1) the following condition is fulfilled

$$\frac{\mu_1\sigma_2 - \mu_2\sigma_1}{\sigma_2\nu_1 - \sigma_1\nu_2} > 0, \quad \sigma_2\nu_1 - \sigma_1\nu_2 \neq 0. \tag{4.21}$$

This measure has a local density

$$Z_t = \left.\frac{d\mathrm{P}^*}{d\mathrm{P}}\right|_{\mathcal{F}_t} = \exp\left\{\alpha^* W_t - \frac{\alpha^{*2}}{2}t + (\lambda - \lambda^*)t + (\ln\lambda^* - \ln\lambda)\Pi_t\right\},$$

where that pair (α^*, λ^*) is the unique solution of the system

$$\left\{\begin{array}{l} \mu_1 = -\sigma_1\alpha^* + \nu_1\lambda^* \\ \mu_2 = -\sigma_2\alpha^* + \nu_2\lambda^* \end{array}\right\}, \tag{4.22}$$

and, therefore

$$\alpha^* = \frac{\mu_2\nu_1 - \mu_1\nu_2}{\sigma_2\nu_1 - \sigma_1\nu_2}, \quad \lambda^* = \frac{\mu_1\sigma_2 - \mu_2\sigma_1}{\sigma_2\nu_1 - \sigma_1\nu_2}.$$

Processes $W_t^* = W_t - \alpha^* t$ and Π_t are independent Wiener and Poisson processes, (with another intensity $\lambda^* > 0$), under the measure P*. As before, in such a market any nonnegative $\mathcal{F}_T$-measurable random variable H will be called a contingent claim.

Let us repeat some definitions and facts for the market (4.20).

Let us take an admissible strategy π and form its value starting at an initial capital $x = X_0^\pi$, bounded by X_0. We call $A(x, \pi)$ the set of successful hedging if

$$A(x, \pi) = \{\omega : X_T^\pi \geqslant H\}.$$

It follows from option pricing theory for complete markets that there exists a strategy π^* with property

$$\mathrm{P}\left(A(\mathrm{E}^*[H], \pi^*)\right) = 1, \tag{4.23}$$

where $X_0^\pi = X_0 = \mathrm{E}^*[H]$, and π^* is a perfect hedge.

However, if $X_0 < \mathrm{E}^*[H]$ then we cannot provide appropriate financing for the perfect hedge in the sense (4.23). In this case, a strategy close enough to the payoff H at maturity in some probabilistic sense is sought. As we already know, one possible solution is to maximize the probability of the successful hedging set over all admissible strategies.

The set $A^* = A\left(X_0^{\pi^*}, \pi^*\right)$ such that

$$\mathrm{P}\{A^*\} = \max_{\pi : X_0^\pi \leqslant X_0} \mathrm{P}\{A(X_0^\pi, \pi)\} \tag{4.24}$$

is called a maximal set for successful hedging. The corresponding strategy π^* is called a quantile hedge.

According to Paragraph 2.1 the optimal (for problem (4.24)) portfolio π^* and the corresponding set A^* can be constructed using arguments of the Neyman–Pearson lemma which leads to the following structure of A^*:

Consider a contingent claim with the payoff H at maturity. Let A^* be a maximal set for successful hedging satisfying (4.24). Then, the structure of this set is

$$A^* = \{Z_T^{-1} \geq a \cdot H\}, \qquad (4.25)$$

and the quantile hedge π^* exists, is unique, and becomes a perfect hedge for the modified claim

$$H_{A^*} = H \mathbb{1}_{A^*}. \qquad (4.26)$$

The maximization problem in (4.24) and the structure of A^* in (4.25) creates a base for a hedging methodology called quantile hedging.

The initial capital bound X_0 is important as a budget restriction for any admissible strategy. According to (4.26), we define the following proposition

$$p = \frac{p\mathrm{E}^* H}{\mathrm{E}^* H} = \frac{X_0}{\mathrm{E}^* H} = \frac{\mathrm{E}^* H_{A^*}}{\mathrm{E}^* H}, \qquad (4.27)$$

as an equivalent characteristic of X_0. The right-hand side of (4.27) is called hedging or loss ratio. We give the full construction of the set A^* for $H = (S_T^1 - S_T^2)^+$, and, keeping in mind further life insurance applications with $p = {}_Tp_x$, we formulate our results in the form (4.27).

The following lemma is a generalization of the Margrabe formula (2.53) to a jump-diffusion model (4.20).

Lemma 4.1. *Consider a financial market with two assets S^1 and S^2 described by jump-diffusion processes (4.20). Then the price of the European option to exchange S^1 for S^2 at time T is given by the following formula*

$$\mathrm{E}^*(S_t^1 - S_T^2)^+ = \sum_{n=0}^{\infty} \mathbb{C}^{Mar}(S_0^1 \vartheta_{n,T}^1, S_0^2 \vartheta_{n,T}^2, T) p_{n,T}^*, \qquad (4.28)$$

where $\mathbb{C}^{Mar}$ denotes a variant of the Margrabe formula: $n = 0, 1, 2, \ldots$

$$\mathbb{C}^{Mar}(\tilde{S}_{0,n}^1, \tilde{S}_{0,n}^2, T) = \tilde{S}_0^1 \Phi(b_+(\tilde{S}_{0,n}^1, \tilde{S}_{0,n}^2, T)) - \tilde{S}_0^2 \Phi(b_-(\tilde{S}_{0,n}^1, \tilde{S}_{0,n}^2, T)),$$

$$\tilde{S}_{0,n}^i = S_0^i \vartheta_{n,T}^i = S_0^i (1-\nu_i)^n \exp(\nu_i \lambda^* T),$$

$$b_{\pm}(\tilde{S}_{0,n}^1, \tilde{S}_{0,n}^2, T)) = \frac{\ln \frac{\tilde{S}_{0,n}^1}{\tilde{S}_{0,n}^2} \pm (\sigma_1 - \sigma_2)^2 \frac{T}{2}}{(\sigma_1 - \sigma_2)\sqrt{T}},$$

$$\Phi(x) = \frac{1}{\sqrt{2\pi}} \int_{-\infty}^{x} \exp(-y^2/2) dy.$$

Proof. First we represent S_t^i, $i = 1, 2$ in an exponential form:

$$\begin{aligned}S_t^i &= S_0^i \exp\left(\sigma_i W_t + \left[\mu_i - \frac{1}{2}(\sigma_i)^2\right]t + \Pi_t \ln(1 - \nu_i)\right) \\ &= S_0^i \exp\left(\sigma_i W_t^* + \left[\nu_i \lambda^* - \frac{1}{2}(\sigma_i)^2\right]t + \Pi_t \ln(1 - \nu_i)\right).\end{aligned} \quad (4.29)$$

Next we consider $H = (S_T^1 - S_T^2)^+$ and find its expected value with respect to P*. Using (4.29) and the independence of W^* and Π under P*, we have

$$\begin{aligned}E^*(S_T^1 - S_T^2)^+ &= E^*\left[E^*\left((S_T^1 - S_T^2)^+ | \Pi_T\right)\right] \\ &= \sum_{n=0}^{\infty} E^*\left[(S_T^1 - S_T^2)^+ | \Pi_T = n\right] p_{n,T}^*,\end{aligned} \quad (4.30)$$

where $p_{n,T}^* = \exp(-\lambda^* T)\frac{(\lambda^* T)^n}{n!}$ are components of the Poisson distribution with intensity $\lambda^* T$.

Note that

$$E^*\left[(S_T^1 - S_T^2)^+ | \Pi_T = n\right] = E^*\left[(s_{n,T}^1 - s_{n,T}^2)^+\right], \quad (4.31)$$

where $s_{n,T}^i$, $i = 1, 2$ are lognormally distributed random variables under P*:

$$\ln s_{n,T}^i \sim N\left(\ln S_0^i(1 - \nu_i)^n + \left[\nu_i \lambda^* - \frac{1}{2}(\sigma_i)^2\right]T, (\sigma_i)^2 T\right), \quad i = 1, 2.$$

Since $\sigma_1 > \sigma_2$, we can apply the Margrabe formula (2.53) in this situation and from (4.30)–(4.31) arrive at (4.28). □

Let us consider the option $(S_t^1 - S_T^2)^+$, where S^1 and S^2 follow a jump diffusion model (4.20). Taking into account the description of quantile hedging methodology given in (4.25)–(4.26) we can rewrite the key relation (4.27) as follows

$$p = \frac{E^*(S_t^1 - S_T^2)^+ \cdot \mathbb{1}_{A^*}}{E^*(S_t^1 - S_T^2)^+}, \quad (4.32)$$

where A^* is the maximal set of successful hedging for $(S_t^1 - S_T^2)^+$.

Let us build the set A^* for the option $(S_t^1 - S_T^2)^+$. To do this, we rewrite the key representation (4.25) for A^* as follows:

$$\begin{aligned}A^* &= \{Z_T^{-1} \geq a \cdot (S_t^1 - S_T^2)^+\} \\ &= \left\{(Z_T S_T^2)^{-1} \geq a \cdot \left(\frac{S_t^1}{S_T^2} - 1\right)^+\right\},\end{aligned} \quad (4.33)$$

where a is an appropriate constant, which will be determined by the initial budget constraint.

It follows from (4.33) that we can work with the ratio $Y_T = \frac{S_T^1}{S_T^2}$.

Using (4.29), we obtain the following exponential form for Y_T:

$$Y_T = \frac{S_0^1}{S_0^2}\left(\frac{1-\nu_1}{1-\nu_2}\right)^{\Pi_T} \exp\left\{(\sigma_1-\sigma_2)W_T + \left[(\mu_1-\mu_2) - \frac{\sigma_1^2-\sigma_2^2}{2}\right]T\right\}$$

$$= \frac{S_0^1}{S_0^2}\left(\frac{1-\nu_1}{1-\nu_2}\right)^{\Pi_T} \exp\left\{(\sigma_1-\sigma_2)W_T^* + \left[(\nu_1-\nu_2)\lambda^* - \frac{\sigma_1^2-\sigma_2^2}{2}\right]T\right\}.$$
(4.34)

Taking into account the formula for Z_T and (4.33)–(4.34), we wish to express A^* in terms of Y_T by rewriting W_T in the following way:

$$W_T = 2W_T - W_t = 2\sigma_1^{-1}(\sigma_1 W_T) - \sigma_2^{-1}(\sigma_2 W_T)$$

$$= 2\sigma_1^{-1}\left[\sigma_1 W_T + \left(\mu_1 - \frac{\sigma_1^2}{2}\right)T\right] - \sigma_2^{-1}\left[\sigma_2 W_T + \left(\mu_2 - \frac{\sigma_2^2}{2}\right)T\right]$$
(4.35)

$$-2\sigma_1^{-1}\left(\mu_1 - \frac{\sigma_1^2}{2}\right)T + \sigma_2^{-1}\left(\mu_2 - \frac{\sigma_2^2}{2}\right)T.$$

Using (4.35), we obtain

$$Z_T S_T^2 = \exp\left\{\alpha^* W_T - \frac{(\alpha^*)^2}{2}T + (\lambda-\lambda^*)T + \Pi_T \ln\frac{\lambda^*}{\lambda}\right\} \cdot S_T^2$$

$$= (S_0^1)^{\frac{2\alpha^*}{\sigma_1}} \exp\left\{\frac{2\alpha^*}{\sigma_1}\left[\sigma_1 W_T + \left(\mu_1 - \frac{\sigma_1^2}{2}\right)T + \Pi_T \ln(1-\nu_1)\right]\right\} \times$$

$$\times (S_0^2)^{-\frac{\alpha^*}{\sigma_2}} \exp\left\{-\frac{\alpha^*}{\sigma_2}\left[\sigma_2 W_T + \left(\mu_2 - \frac{\sigma_2^2}{2}\right)T + \Pi_T \ln(1-\nu_2)\right]\right\} \times$$

$$\times S_T^2 (S_0^1)^{-\frac{2\alpha^*}{\sigma_1}}(S_0^2)^{\frac{\alpha^*}{\sigma_2}} \exp\left\{\Pi_T \ln\left[(1-\nu_1)^{-\frac{2\alpha^*}{\sigma_1}}(1-\nu_2)^{\frac{\alpha^*}{\sigma_2}}\frac{\lambda^*}{\lambda}\right]\right\} \times$$

$$\times \exp\left\{-\frac{2\alpha^*}{\sigma_1}\left(\mu_1 - \frac{\sigma_1^2}{2}\right)T + \frac{\alpha^*}{\sigma_2}\left(\mu_2 - \frac{\sigma_2^2}{2}\right)T - \frac{(\alpha^*)^2}{2}T - (\lambda-\lambda^*)T\right\}$$

$$= (S_T^1)^{\frac{2\alpha^*}{\sigma_1}}(S_T^2)^{1-\frac{\alpha^*}{\sigma_2}} \cdot b^{\Pi_T} \cdot g,$$
(4.36)

where

$$b = (1-\nu_1)^{-\frac{2\alpha^*}{\sigma_1}}(1-\nu_2)^{\frac{\alpha^*}{\sigma_2}}\frac{\lambda^*}{\lambda},$$

$$g = (S_0^1)^{-\frac{2\alpha^*}{\sigma_1}}(S_0^2)^{\frac{\alpha^*}{\sigma_2}} \exp\left\{-\frac{2\alpha^*}{\sigma_1}\left(\mu_1 - \frac{\sigma_1^2}{2}\right)T\right.$$

$$\left.+ \frac{\alpha^*}{\sigma_2}\left(\mu_2 - \frac{\sigma_2^2}{2}\right)T - \frac{(\alpha^*)^2}{2}T - (\lambda-\lambda^*)T\right\}.$$

Let us write (4.36) in the form

$$Z_T S_T^2 = (Y_T)^\alpha \cdot b^{\Pi_T} \cdot g,$$
(4.37)

where the work parameter α is chosen as

$$\frac{2\alpha^*}{\sigma_1} = \alpha = \frac{\alpha^*}{\sigma_2} - 1.$$

Hence,

$$\alpha^* = \frac{\sigma_1 \sigma_2}{\sigma_1 - 2\sigma_2}, \quad \sigma_1 \neq 2\sigma_2. \tag{4.38}$$

Taking into account (4.22), (4.37) and (4.38), we arrive at the following condition for the parameters of the model (4.20):

$$\frac{\mu_2 \nu_1 - \mu_1 \nu_2}{\sigma_2 \nu_1 - \sigma_1 \nu_2} = \frac{\sigma_1 \sigma_2}{\sigma_1 - 2\sigma_2}, \tag{4.39}$$

where $\sigma_1 > \sigma_2$, $\sigma_1 \neq 2\sigma_2$, and $\sigma_2 \nu_1 \neq \sigma_1 \nu_2$.

Relations (4.33) and (4.37) give

$$A^* = \left\{ Y_T^{-\frac{2\alpha^*}{\sigma_1}} \geq b^{\Pi_T} \cdot g \cdot a(Y_T - 1)^+ \right\}. \tag{4.40}$$

To analyze A^* represented in the form (4.40), we consider the set $\{\Pi_T = n\}$, $n = 0, 1, 2, \ldots$ and the following characteristic equation

$$x^{-\frac{2\alpha^*}{\sigma_1}} = b^n \cdot g \cdot a(x-1)^+. \tag{4.41}$$

Assuming (4.39), we distinguish two cases:

Case 1: $\sigma_1 > 2\sigma_2$ (or $-\frac{2\alpha^*}{\sigma_1} \leq 1$) and

Case 2: $\sigma_2 < \sigma_1 < 2\sigma_2$ (or $-\frac{2\alpha^*}{\sigma_1} > 1$).

Exploiting (4.38), we can easily check that (4.41) admits the only solution $c(n)$ in the first case and two solutions $c_1(n) < c_2(n)$ in the second case. Hence, we shall use them to construct the set of successful hedging in the form $\{Y_T \leq c\}$ for the first case and $\{Y_T \leq c_1\} \cup \{Y_T > c_2\}$ for the second case on each set $\{\Pi_T = n\}$, $n = 0, 1, \ldots$.

Now we are ready to formulate the main result from the application of quantile hedging to the option to exchange one asset for another.

Theorem 4.2. *For an option $(S_T^1 - S_T^2)^+$, where S^1 and S^2 follow a jump-diffusion model (4.20), $\mu_1 > \mu_2$, $\sigma_1 > \sigma_2$, and (4.21) and (4.39) are fulfilled, the following formulae for the loss ration (4.27) are true:*

1. *If $\sigma_1 > 2\sigma_2$ then*

$$p = 1 - \frac{\sum_0^\infty p_{n,T}^* \left[\tilde{S}_{0,n}^1 \Phi(b_+(\tilde{S}_{0,n}^1, c\tilde{S}_{0,n}^2, T)) - \tilde{S}_{0,n}^2 \Phi(b_-(\tilde{S}_{0,n}^1, c\tilde{S}_{0,n}^2, T)) \right]}{\sum_0^\infty p_{n,T}^* \left[\tilde{S}_{0,n}^1 \Phi(b_+(\tilde{S}_{0,n}^1, \tilde{S}_{0,n}^2, T)) - \tilde{S}_{0,n}^2 \Phi(b_-(\tilde{S}_{0,n}^1, \tilde{S}_{0,n}^2, T)) \right]},$$

(4.42)

where $c = c(n)$ is a unique solution to the equation (4.41).

2. If $\sigma_2 < \sigma_1 < 2\sigma_2$ then

$$p = 1 - \frac{\sum_0^\infty p_{n,T}^* \left[\tilde{S}_{0,n}^1 \Phi(b_+(\tilde{S}_{0,n}^1, c_1\tilde{S}_{0,n}^2, T)) - \tilde{S}_{0,n}^2 \Phi(b_-(\tilde{S}_{0,n}^1, c_1\tilde{S}_{0,n}^2, T))\right]}{\sum_0^\infty p_{n,T}^* \left[\tilde{S}_{0,n}^1 \Phi(b_+(\tilde{S}_{0,n}^1, \tilde{S}_{0,n}^2, T)) - \tilde{S}_{0,n}^2 \Phi(b_-(\tilde{S}_{0,n}^1, \tilde{S}_{0,n}^2, T))\right]}$$

$$+ \frac{\sum_0^\infty p_{n,T}^* \left[\tilde{S}_{0,n}^1 \Phi(b_+(\tilde{S}_{0,n}^1, c_2\tilde{S}_{0,n}^2, T)) - \tilde{S}_{0,n}^2 \Phi(b_-(\tilde{S}_{0,n}^1, c_2\tilde{S}_{0,n}^2, T))\right]}{\sum_0^\infty p_{n,T}^* \left[\tilde{S}_{0,n}^1 \Phi(b_+(\tilde{S}_{0,n}^1, \tilde{S}_{0,n}^2, T)) - \tilde{S}_{0,n}^2 \Phi(b_-(\tilde{S}_{0,n}^1, \tilde{S}_{0,n}^2, T))\right]},$$
(4.43)

where $c_1 < c_2$ are solutions to the equation (4.41), and $p_{n,T}^* = \exp(-\lambda^* T)\frac{(\lambda^* T)^n}{n!}$ are components of a Poisson distribution with intensity $\lambda^* T$,

$$\lambda^* = \frac{\mu_1\sigma_2 - \mu_2\sigma_1}{\sigma_2\nu_1 - \sigma_1\nu_2},$$

$$\tilde{S}_{0,n}^i = S_0^i(1-\nu_i)^n \exp(\nu_i\lambda^* T),$$

$$b_\pm(\tilde{S}_{0,n}^1, \tilde{S}_{0,n}^2, T) = \frac{\ln\frac{\tilde{S}_{0,n}^1}{\tilde{S}_{0,n}^2} \pm (\sigma_1 - \sigma_2)^2\frac{T}{2}}{(\sigma_1-\sigma_2)\sqrt{T}},$$

$$\Phi(x) = \frac{1}{\sqrt{2\pi}}\int_{-\infty}^x \exp(-y^2/2)dy.$$

Proof. Since the denominator $E^*(S_T^1 - S_T^2)^+$ of (4.32) is given in Lemma 4.1, we will work with the numerator $E^*(S_T^1 - S_T^2)^+ \mathbb{1}_{A^*}$. It is sufficient to compute

$$E^*\left[(S_T^1 - S_T^2)^+ \mathbb{1}_{A^*} | \Pi_T = n\right],$$

with further averaging over the Poisson distribution $p_{n,T}^*$. We consider two cases depending on whether (4.41) has one or two solutions separately.

Case 1: $\sigma_1 > 2\sigma_2$ (or $-\frac{2\alpha^*}{\sigma_1} \leq 1$)
As $c(n) > 1$, it is easy to see that

$$E^*\left[(S_T^1 - S_T^2)^+ \mathbb{1}_{\{Y_T \leq c(n)\}}|\Pi_T = n\right] = E^*\left[(S_T^1 - S_T^2)^+|\Pi_T = n\right]$$
$$-E^*\left[S_T^1 \mathbb{1}_{\{Y_T > c(n)\}}|\Pi_T = n\right] + E^*\left[S_T^2 \mathbb{1}_{\{Y_T > c(n)\}}|\Pi_T = n\right] \quad (4.44)$$
$$= E^*\left[(S_T^1 - S_T^2)^+|\Pi_T = n\right] - E^*\left[(S_T^1 - S_T^2)|\Pi_T = n\right]$$
$$+E^*\left[S_T^1 \mathbb{1}_{\{Y_T \leq c(n)\}}|\Pi_T = n\right] - E^*\left[S_T^2 \mathbb{1}_{\{Y_T \leq c(n)\}}|\Pi_T = n\right]$$

The first term in the last equality of (4.44) is given by (4.31). To find the second term, we use (4.29) and obtain

$$E^*\left[(S_T^1 - S_T^2)|\Pi_T = n\right]$$
$$= \exp\left\{\ln S_0^1(1-\nu_1)^n + \nu_1\lambda^* T\right\} - \exp\left\{\ln S_0^1(1-\nu_2)^n + \nu_2\lambda^* T\right\} \quad (4.45)$$
$$= \tilde{s}_{0,n}^1 - \tilde{s}_{0,n}^2.$$

To calculate expectations
$$E^*[S_T^i \mathbb{1}_{\{Y_T \leq c(n)|\Pi_T=n\}}], \quad i=1,2, \tag{4.46}$$
we rewrite
$$\{Y_T \leq c(n)\} = \{\ln Y_T \leq \ln c(n)\}$$
and denote $\xi = \ln Y_T$, $S_T^i = \exp\{-\eta_i\}$, where the Gaussian random variables
$$\eta_i = -\left[\ln\left(S_0^i(1-\nu_i)^n\right) + \sigma_i W_T^* + \left(\nu_i \lambda^* - \frac{\sigma_i^2}{2}\right)T\right], \quad i=1,2,$$
are defined by (4.29). Under the condition $\{\Pi_T = n\}$, the pairs (ξ, η_1) and (ξ, η_2) are two systems of Gaussian random variables. According to Lemma 2.3, for $i=1,2$ we obtain
$$E^*\left[\exp\{-\eta_i\}\mathbb{1}_{\{\xi \leq \ln c\}}|\Pi_T = n\right]$$
$$= \exp\left\{\frac{\sigma_{\eta_i}^2}{2} - \mu_{\eta_i}\right\} \Phi\left(\frac{\ln(c) - (\mu_\xi - \text{cov}(\xi, \eta_i))}{\sigma_\xi}\right).$$

The parameters μ and σ^2 can be easily determined from (4.29) and (4.34):
$$\mu_\xi = \mu_{\ln Y_T} = E^*\left[\ln(Y_T)|\Pi_T = n\right]$$
$$= \ln\left\{\frac{S_0^1}{S_0^2}\left(\frac{1-\nu_1}{1-\nu_2}\right)^n\right\} + \left[(\nu_1 - \nu_2)\lambda^* - \frac{\sigma_1^2 - \sigma_2^2}{2}\right]T,$$
$$\sigma_\xi^2 = (\sigma_1 - \sigma_2)^2 T,$$
$$\mu_{\eta_i} = E^*[\eta_i|\Pi_T = n] = -\ln S_0^1(1-\nu_1)^n - \left[\nu_i\lambda^* - \frac{\sigma_i^2}{2}\right]T,$$
$$\sigma_{\eta_i}^2 = \sigma_i^2 T,$$
$$\text{cov}(\xi, \eta_i) = -\sigma_i(\sigma_1 - \sigma_2)T, \quad i = 1,2.$$

Putting these values into the formula (4.44), we find the values of the expectations in (4.45) and also the difference of the last terms of (4.44)
$$E^*\left[(S_T^1 - S_T^2)\mathbb{1}_{\{Y_T \leq c(n)\}}|\Pi_T = n\right]$$
$$= \tilde{S}_{0,n}^1 \Phi\left(\frac{\ln c(n)}{(\sigma_1 - \sigma_2)\sqrt{T}} - b_+\left(\tilde{S}_{0,n}^1, \tilde{S}_{0,n}^2, T\right)\right)$$
$$- \tilde{S}_{0,n}^2 \Phi\left(\frac{\ln c(n)}{(\sigma_1 - \sigma_2)\sqrt{T}} - b_-\left(\tilde{S}_{0,n}^1, \tilde{S}_{0,n}^2, T\right)\right) \tag{4.47}$$
$$= \left(\tilde{S}_{0,n}^1 - \tilde{S}_{0,n}^2\right) - \left[\tilde{S}_{0,n}^1 \Phi\left(b_+\left(\tilde{S}_{0,n}^1, c\tilde{S}_{0,n}^2, T\right)\right)\right.$$
$$\left. - \tilde{S}_{0,n}^2 \Phi\left(b_-\left(\tilde{S}_{0,n}^1, c\tilde{S}_{0,n}^2, T\right)\right)\right].$$

Combining (4.44), (4.45) and (4.47) with (4.32), we finally arrive at (4.42).

Case 2: $\sigma_2 < \sigma_1 < 2\sigma_2$ (or $-\frac{2\alpha^*}{\sigma_1} > 1$)

The second case can be treated in a similar way. The set of successful hedging A^*, (again on $\{\Pi_T = n\}$, $n = 0, 1, \ldots$), consists of two parts: $\{Y_T \leqslant c_1(n)\}$ and $\{Y_T > c_2(n)\}$. Hence, we have

$$\mathbb{1}_{A^*} = \mathbb{1}_{\{Y_T \leqslant c_1\} \cup \{Y_T > c_2(n)\}}. \qquad (4.48)$$

Using the inequalities $1 \leqslant c_1 \leqslant c_2$ and (4.48), we obtain

$$\begin{aligned} & \mathrm{E}^* \left[\mathbb{1}_{A^*} (S_T^1 - S_T^2)^+ | \Pi_T = n \right] \\ & = \mathrm{E}^* \left[(S_T^1 - S_T^2)^+ | \Pi_T = n \right] + \mathrm{E}^* \left[\mathbb{1}_{\{Y_T \leqslant c_1\}} (S_T^1 - S_T^2) | \Pi_T = n \right] \qquad (4.49) \\ & \quad - \mathrm{E}^* \left[\mathbb{1}_{\{Y_T \leqslant c_2\}} (S_T^1 - S_T^2) | \Pi_T = n \right]. \end{aligned}$$

The first term in the right-hand side of (4.49) is calculated with the help of (4.31). The other two terms in (4.49) are found as in (4.46) with evident changes. All these calculations lead us to the explicit form for (4.32) in the second case given in (4.43). □

Remark 4.1. Under known proportion p, formulae (4.42) and (4.43) provide the possibility to determine the maximal successful hedging set A^*. The corresponding hedge π^* will be a perfect hedge for a modified claim (4.26). Its capital $\mathbb{C}(t, S_t^1, S_t^2)$ can be computed in a way similar to the initial price of the option.

Remark 4.2. We can fix the probability of the set of successful hedging as $1 - \epsilon$, $\epsilon > 0$. Then, applying (4.42) or (4.43), we can find the proportion p and identify the corresponding initial capital from this risk level.

4.2 Quantile hedging and an exemplary risk-management scheme for equity-linked life insurance

The basic goal of this paragraph is to collect together three market models, the Black–Scholes model (Model 1), two-factor diffusion (Model 2) and jump-diffusion models (Model 3), and two types of equity-linked life insurance contracts with constant and flexible/stochastic guarantees. The general idea behind this is to reflect both their advantages and disadvantages in the sense of quantile hedging. Based on such approach we propose an exemplary scheme for risk-management of equity-linked life insurance contracts. This is demonstrated both, theoretically and numerically, and is closer to a practical implementation of our results. To achieve a better understanding and to provide convenience for the reader, we reproduce here in a concentrated manner the main steps and features of the quantile hedging methodology. Let us start with the financial settings.

Model 1

We consider a financial market consisting of a non-risky asset (bank-account) $B_t = e^{rt}$, $r \geq 0$, $t \geq 0$, and a risky asset which follows the Black–Scholes model

$$dS_t = S_t(\mu dt + \sigma dW_t),$$

where W is a Wiener process, μ and σ are the rate of return and the volatility of the asset S, and the initial investment equals S_0.

The market in this setting is complete and has the unique risk-neutral probability defined by

$$Z_T = \exp\left\{-\frac{\mu - r}{\sigma}W_T - \frac{1}{2}\left(\frac{\mu - r}{\sigma}\right)^2 T\right\}.$$

Model 2

A financial market consists of a non-risky asset $B_t = e^{rt}$, $t \geq 0$, $r \geq 0$ and two risky assets S^1 and S^2.

The risky assets follow the two-dimensional Black–Scholes model

$$dS_t^i = S_t^i(\mu_i dt + \sigma_i dW_t), \quad i = 1, 2,$$

where W is a Wiener process and μ_i and σ_i are the rate of return and the volatility of the asset S^i.

Both assets have the same source of randomness generated by the same Wiener process. We demand that $\mu_1 > \mu_2$, $\sigma_1 > \sigma_2$ since S^2 is supposed to prove a flexible guarantee and should be less risky than S^1. Also, the initial values of both assets should be equal to $S_0^1 = S_0^2 = S_0$ and are considered as the initial investment.

In order to prevent the existence of arbitrage opportunities, we suppose that the following technical condition is fulfilled

$$\frac{\mu_1 - r}{\sigma_1} = \frac{\mu_2 - r}{\sigma_2}. \tag{4.50}$$

Then the risk-neutral probability P* has the following density with respect to the initial probability

$$Z_T = \exp\left\{-\frac{\mu_1 - r}{\sigma_1}W_T - \frac{1}{2}\left(\frac{\mu_1 - r}{\sigma_1}\right)^2 T\right\}.$$

Model 3

As in Model 2, we consider a financial market consisting of a non-risky asset $B_t = e^{rt}$, $t \geq 0$, $r \geq 0$ and two risky assets S^1 and S^2. But now the risky assets follow a jump-diffusion model

$$dS_t^i = S_{t-}^i(\mu_i dt + \sigma_i dW_t - \nu_i d\Pi_t), \quad i = 1, 2,$$

where W is a Wiener process, Π is a Poisson process with a positive intensity λ, ν_i describes the size of the jumps for S^i, and the following conditions on the parameters of the model should be satisfied

$$\mu_1 > \mu_2, \quad \sigma_1 > \sigma_2, \quad \nu_1, \nu_2 < 1.$$

$S_0^1 = S_0^2 = S_0$ is the initial investment.

In order to prevent the existence of arbitrage opportunities, we demand that

$$\sigma_2 \nu_1 - \sigma_1 \nu_2 \neq 0,$$
$$\frac{(\mu_1 - r)\sigma_2 - (\mu_2 - r)\sigma_1}{\sigma_2 \nu_1 - \sigma_1 \nu_2} > 0. \tag{4.51}$$

Then, a risk-neutral probability will have the following density with respect to the initial probability

$$Z_t = \exp\left\{\alpha^* W_T - \frac{(\alpha^*)^2}{2} T + (\lambda - \lambda^*)t + \ln\frac{\lambda^*}{\lambda} \cdot \Pi_t\right\},$$

where (α^*, λ^*) are the unique solutions to the system of equations

$$\begin{cases} \mu_1 - r = -\sigma_1 \alpha^* + \nu_1 \lambda^*, \\ \mu_2 - r = -\sigma_2 \alpha^* + \nu_2 \lambda^*, \quad \lambda^* > 0. \end{cases} \tag{4.52}$$

Parameter λ^* has a good probabilistic interpretation as a new intensity of the Poisson process with respect to the risk-neutral probability P^*.

Assume that all models are given on a standard stochastic basis $(\Omega, \mathcal{F}, \mathbb{F} = (\mathcal{F}_t)_{t \geq 0}, P)$.

For Model 1, let $\pi_t = (\beta_t, \gamma_t)$ be a portfolio consisting of β_t units of the non-risky asset and γ_t units of the risky asset. The portfolio must be admissible, self-financing and adapted to the filtration $\mathbb{F}$. Then, the value of the portfolio at any time is equal to $X_t^\pi = \beta_t B_t + \gamma_t S_t$. For Models 2 and 3, a portfolio $\pi_t = (\beta_t, \gamma_t^1, \gamma_t^2)$ consists of β_t units of the non-risky asset and γ_t^1, γ_t^2 units of the risky assets S^1 and S^2 and has the value $X_t^\pi = \beta_t B_t + \gamma_t^1 S_t^1 + \gamma_t^2 S_t^2$.

Every $\mathcal{F}_T$-measurable nonnegative random variable H is called a contingent claim. A self-financing strategy π is a perfect hedge for H if $X_T^\pi \geq H$ (a.s.). According to the option pricing theory, it does exist, is unique for a given contingent claim, and has the initial value $X_0^\pi = e^{-rT} E^* H$.

Further, we consider equity-linked pure endowment contracts with guarantees. An insurance risk to which the insurance company is exposed when it enters into such contracts includes two components. The first one is based on the survival of a client at maturity as at that time the insurance company would be obliged to pay a benefit to the living insured. We call it a mortality risk. The second component depends on a mortality frequency risk for a pooled number of similar contracts. A large enough portfolio of life insurance contracts will result in a more predictable mortality risk exposure and a reduced mortality frequency risk. We will work first with the mortality risk of a single contract. We will show later how the mortality frequency risk can be treated.

We use a random variable $T(x)$ on a probability space $(\tilde{\Omega}, \tilde{\mathcal{F}}, \tilde{P})$ to denote the remaining lifetime of a person of age x. Let $_T p_x = \tilde{P}\{T(x) > T\}$ be a survival probability for the next T years of the insured. It is reasonable to assume that $T(x)$ does not depend on the evolution of the financial market.

We study contracts with fixed and flexible/stochastic guarantees which make a payment at maturity provided the insured is alive. The benefit will depend on the evolution of a financial market. Due to independency of the "financial" and the "insurance" parts of the contracts, we consider them on the product probability space $(\Omega \times \tilde{\Omega}, \mathcal{F} \times \tilde{\mathcal{F}}, P \times \tilde{P})$. The benefit of the contracts with a fixed guarantee is linked to the prices of a financial market described by Model 1 and is equal to

$$H(T(x)) = \max\{S_T, k \cdot S_0\} \cdot \mathbb{1}_{\{T(x) > T\}}, \qquad (4.53)$$

where k is a fixed coefficient equal to a percentage of the initial contribution guaranteed to the living insured at maturity. The contracts with a flexible guarantee are considered in the framework of Models 2 and 3. In this case the payoff at maturity is

$$H(T(x)) = \max\{S_T^1, S_T^2\} \cdot \mathbb{1}_{\{T(x) > T\}}. \qquad (4.54)$$

A strategy with payoff $H = \max\{S_T, k \cdot S_0\}$ at maturity is a perfect hedge for contract (4.53), whereas a strategy with payoff $H = \max\{S_T^1, S_T^2\}$ is a perfect hedge for contract (4.54). Their prices are equal to the $E^* H e^{-rT}$.

Let us rewrite the financial components of (4.53) and (4.54) as follows

$$H = \max\{S_T, k \cdot S_0\} = kS_0 + (S_T - kS_0)^+ \text{ for } (4.53), \qquad (4.55)$$

$$H = \max\{S_T^1, S_T^2\} = S_T^2 + (S_T^1 - S_T^2)^+, \text{ for } (4.54). \qquad (4.56)$$

Using (4.55) and (4.56), we reduce the pricing of the claims (4.53) and (4.54) to the pricing of embedded options $(S_T - kS_0)^+$ and $(S_T^1 - S_T^2)^+$, respectively, provided $\{T(x) > T\}$.

According to the well-developed option pricing theory, the optimal price is traditionally calculated as an expected present value of cash flows under the risk-neutral probability. However, the "insurance" part of contracts (4.53) and (4.54) does not need to be risk-adjusted since the mortality risk is essentially unsystematic. It means that the mortality risk can be effectively eliminated by increasing the number of similar insurance policies.

Recall the so-called Brennan–Schwartz price for contracts (4.53) and (4.54) is equal to

$$_T U_x = E^* \times \tilde{E} H(T(x)) e^{-rT} = {_T p_x} k S_0 e^{-rT} + {_T p_x} E^*(S_T - kS_0)^+ e^{-rT}, \quad (4.57)$$

$$_T U_x = E^* \times \tilde{E} H(T(x)) = {_T p_x} S_0 + {_T p_x} E^*(S_T^1 - S_T^2)^+ e^{-rT} \qquad (4.58)$$

correspondingly, where $E^* \times \tilde{E}$ is the expectation with respect to $P^* \times \tilde{P}$.

The insurance company acts as a hedger of H in the financial market. It follows from (4.57) and (4.58) that the initial price of H is strictly less than that of the perfect hedge, since a survival probability is always less than one, or

$$_T U_x < E^* H e^{-rT}.$$

Therefore, perfect hedging of H with an initial value of the hedge restricted by the $E^* H e^{-rT}$ is not possible. Instead, we will look for a strategy π^* with some initial budget constraint such that its value $X_T^{\pi^*}$ at maturity is close to H in some probabilistic sense. Quantile hedging presents a methodology to realize this idea.

To find the current price H_t of a contingent claim H at any time before maturity T, we would like to construct a portfolio π^* such that $X_T^{\pi^*}$ is close enough to H. For example, for a market with the unique risk-neutral probability P^* the price H_t can be calculated as follows

$$H_t = E^*(H|\mathcal{F}_t)e^{-r(T-t)} \text{ and } P(X_T^{\pi^*}(H_0) = H) = 1,$$

where $H_0 = E^* H e^{-rT}$ is an initial capital invested in the perfect hedge π^*.

Now assume that there is a budget constraint imposed on the portfolio π

$$X_0^\pi \leqslant X_0 < H_0,$$

which means that at initial time $t = 0$ we are able to invest in π less than H_0. Therefore, we cannot provide perfect hedging of the claim H. Instead, we would like to maximize the probability that at maturity the value of the portfolio π will be at least equal to the claim H. This probability plays the role of a risk measure naturally adapted to the problem under our consideration.

Going in this way we define the following set $A(X_0^\pi, \pi) = \{\omega : X_T^\pi(X_0^\pi) \geqslant H\}$. It represents all possible states of the financial market under which investing a restricted budget X_0^π in the portfolio π at the initial time enables us to pay off the obligation H at maturity. We will call it a successful hedging set. It depends on both the initial budget and the portfolio chosen.

The portfolio π^* is called a quantile hedge if the probability of a successful hedging set for it is maximal

$$P\left\{A\left(X_0^{\pi^*}, \pi^*\right)\right\} = \max_{\pi: X_0^\pi \leqslant X_0} P\left\{A\left(X_0^\pi, \pi\right)\right\}.$$

How can the optimal (in this sense) portfolio π^* and the corresponding set $A\left(X_0^{\pi^*}, \pi^*\right)$ be constructed? The answer to this question is given below (see also Chapter 2).

Consider a contingent claim with the payoff H at maturity. Let $A^* \in \mathcal{F}_T$ be a solution to the extreme problem

$$P(A^*) = \max_{A \in \mathcal{F}_T : E^*(H \cdot \mathbb{1}_A) \leqslant X_0} P(A).$$

Then the quantile hedge π^*

- Does exist,
- Is unique,
- Is a perfect hedge for a modified claim $H_{A*} = H \cdot \mathbb{1}_{A*}$.

Moreover, the structure (see, the Neyman–Pearson lemma) of the maximal successful hedging set is $A^* = \{Z_T^{-1} > aH\}$ where Z_T is the density of a risk-neutral probability with respect to the initial probability and a constant a is defined by X_0.

We will apply the quantile hedging methodology to the financial components of the equity-linked contracts (4.53) and (4.54). Using the answer to the above optimization problem, we could reduce a construction of a quantile hedge for the claim H to a construction of a perfect hedge for a modified claim H_{A*}. Once the quantile hedges for (4.55) and (4.56) are constructed, we will set their prices equal to the initial values $X_0^{\pi^*}$ of the corresponding quantile hedges and draw conclusions about the relationships between financial and insurance components of the contracts.

We make an assumption that financial and insurance risks are independent. As a result, we could hedge them independently. To do so we separate financial and insurance risks. As a result of this separation, the financial risk can be minimized by constructing an appropriate financial strategy while the insurance risk is diversified based on the law of large numbers. We do not require perfect hedging and it is not possible. Instead, we apply quantile hedging to the financial component of the contract allowing for the possibility that a client probably will not survive until maturity.

Since (4.55) and (4.56) are true, we will pay attention to the terms $(S_T - kS_0)^+ \mathbb{1}_{\{T(x)>T\}}$ and $(S_T^1 - S_T^2)^+ \mathbb{1}_{\{T(x)>T\}}$ associated with a call option and an option to exchange one asset for another. Consider $_Tp_x \cdot E^*(S_T - kS_0)^+ e^{-rT}$ and $_Tp_x \cdot E^*(S_T^1 - S_T^2)^+ e^{-rT}$ as bounds on the initial capital available for portfolios with the payoffs $\max(0, S_T - kS_0)$ and $\max(0, S_T^1 - S_T^2)$, respectively. Using the definitions of perfect and quantile hedging, we arrive at the following equalities for the contracts with fixed and flexible/stochastic guarantees

$$_Tp_x \cdot E^*(S_T - kS_0)^+ = E^*\left((S_T - kS_0)^+ \cdot \mathbb{1}_{A*}\right),$$

$$_Tp_x \cdot E^*(S_T^1 - S_T^2)^+ = E^*\left((S_T^1 - S_T^2)^+ \cdot \mathbb{1}_{A*}\right).$$

As a result, we can separate insurance and financial components of the contracts as follows

$$_Tp_x = \frac{E^*\left((S_T - kS_0)^+ \cdot \mathbb{1}_{A*}\right)}{E^*(S_T - kS_0)^+}, \qquad (4.59)$$

$$_Tp_x = \frac{E^*\left((S_T^1 - S_T^2)^+ \cdot \mathbb{1}_{A*}\right)}{E^*(S_T^1 - S_T^2)^+}. \qquad (4.60)$$

The left-hand sides of (4.59) and (4.60) are equal to the survival probability of the insured, which is a mortality risk for the insurer, while the right-hand sides are related to a pure financial risk as they are connected to the evolution of the financial market. So, equations (4.59) and (4.60) can be viewed as the key balance equations separating the risks associated with contracts (4.53) and (4.54).

It can be shown that maximal successful hedging sets that appear in (4.59) and (4.60) have the following structures

$$A^* = \{Z_T^{-1} > a(S_T - kS_0)^+\} = \left\{\left(\frac{S_T}{kS_0}\right)^{-q} \geq g \cdot \left(\frac{S_T}{kS_0} - 1\right)^+\right\},$$

$$A^* = \{Z_T^{-1} > a(S_T^1 - S_T^2)^+\} = \left\{\left(\frac{S_T^1}{S_T^2}\right)^{-q} \geq g \cdot \left(\frac{S_T^1}{S_T^2} - 1\right)^+\right\},$$

where g are constants depending on a, T, r, μ, σ (for Model 1) or a, T, r, μ_1, μ_2, σ_1, σ_2 (for Models 2 and 3), and the power q depend on r, μ, σ and r, μ_1, μ_2, σ_1, σ_2, respectively. Note that the characteristic equation

$$y^{-q} = g \cdot (y - 1)^+ \tag{4.61}$$

associated with the above sets has a unique solution if $-q \leq 1$ and two solutions if $-q > 1$. A further analysis relies heavily on the properties of the diffusion and jump-diffusion processes of Models 1, 2 and 3 underlying the evolution of the risky assets.

The following theorems represent the resulting developments of formulae (4.59) and (4.60). Formally, the contracts with a fixed guarantee can be treated as a particular case of Model 2 (or even 3) with $\mu_2 - r = \sigma_2 = 0$. But instead, we prefer to use directly the formulae for the maximal successful hedging set in this case. The proof of Theorem 4.3 can be easily produced as in Paragraph 2.1 and therefore omitted here. The proof of Theorem 4.5 is given at the end of this paragraph. The proof of Theorem 4.4 is analogous.

Theorem 4.3. *Consider a financial market with a risky asset following Model 1 and an equity-linked life insurance contract with a fixed guarantee and payoff* $(S_T - kS_0)^+ \cdot \mathbb{1}_{\{T(x) > T\}}$ *linked to the evolution of the risky asset. Then, it is possible to balance the survival probability of an insured and the financial risk associated with the contract as follows:*

$$_Tp_x = 1 - \frac{\Phi\left(\frac{-b+\sigma T}{\sqrt{T}}\right) - ke^{-rT}\Phi\left(\frac{-b}{\sqrt{T}}\right)}{\Phi(d_+(1,k,T)) - ke^{-rT}\Phi(d_-(1,k,T))} \quad \text{if } \mu - r \leq \sigma^2,$$

$$_Tp_x = 1 - \frac{\Phi\left(\frac{-b_1+\sigma T}{\sqrt{T}}\right) - \Phi\left(\frac{-b_2+\sigma T}{\sqrt{T}}\right) - ke^{-rT}\left[\Phi\left(\frac{-b_1}{\sqrt{T}}\right) - \Phi\left(\frac{-b_2}{\sqrt{T}}\right)\right]}{\Phi(d_+(1,k,T)) - ke^{-rT}\Phi(d_-(1,k,T))}$$

$$\text{if } \mu - r > \sigma^2,$$

$$\tag{4.62}$$

where $d_\pm(x,y,T) = \frac{\ln\frac{x}{y}+(r\pm\frac{1}{2}\sigma^2)\cdot T}{\sigma\cdot\sqrt{T}}$, $\Phi(x)$ is a standard normal distribution, $b = \frac{\ln\tilde{b}k-(r-\frac{1}{2}\sigma^2)\cdot T}{\sigma}$, $\tilde{b}$ is a unique solution to equation (4.61) when $\mu-r \leqslant \sigma^2$, b_1 and b_2 are defined in a similar way using two solutions $\tilde{b}_1$ and $\tilde{b}_2$ to (4.61) for the case when $\mu-r > \sigma^2$.

Note that the results in (4.62) are given in terms of the Black–Scholes formula.

Remark 4.3. The solution (or the solutions) to (4.61) defining a constant b (or constants b_1, b_2) can be found from the structure of a maximal successful hedging set and its probability. To do this, the insurance company chooses some risk level $\epsilon \in (0,1)$ it is ready to take. In other words, ϵ is a probability that the company would not be able to hedge the contingent claim H perfectly or $1-\epsilon = P(A^*)$.

For example, if $\mu - r \leqslant \sigma_2$ then (4.61) has the unique solution $\tilde{b}$ and we represent A^* as $\left\{\frac{S_T}{kS_0} \leqslant \tilde{b}\right\}$. Then $\left\{\frac{S_T}{kS_0} \leqslant \tilde{b}\right\} = 1-\epsilon$ and, using the log-normality of S_T, we arrive at the following equality

$$P(A^*) = \Phi\left(\frac{b-\frac{\mu-r}{\sigma}T}{\sqrt{T}}\right), \quad b = \sqrt{T}\Phi^{-1}(1-\epsilon) + \frac{\mu-r}{\sigma}T.$$

If the characteristic equation (4.61) has two solutions $\tilde{b}_1$ and $\tilde{b}_2$ (the case $\mu - r > \sigma^2$), they can be found from $P\left(\left\{\frac{S_T^1}{kS_0} \leqslant \tilde{b}_1\right\} \cup \left\{\frac{S_T^1}{kS_0} \geqslant \tilde{b}_2\right\}\right) = 1-\epsilon$ in a similar way.

Theorem 4.4. *Consider a financial market with two risky assets following Model 2 and an equity-linked life insurance contract with payoff $(S_T^1 - S_T^2)^+ \cdot \mathbb{1}_{\{T(x)>T\}}$ linked to the evolution of the risky assets. We assume that (4.50) is true. Then, it is possible to balance the survival probability of an insured and the financial risk associated with the contract as follows:*

$$_T p_x = 1 - \frac{\Phi(b_+(1,c,T)) - \Phi(b_-(1,c,T))}{\Phi(b_+(1,1,T)) - \Phi(b_-(1,1,T))} \quad \text{if } -q \leqslant 1,$$

$$_T p_x = 1 - \frac{\Phi(b_+(1,c_1,T)) - \Phi(b_-(1,c_1,T)) + \Phi(b_+(1,c_2,T)) - \Phi(b_-(1,c_2,T))}{\Phi(b_+(1,1,T)) - \Phi(b_-(1,1,T))}$$

$$\text{if } -q > 1,$$

(4.63)

where $b_\pm(x,y,T) = \frac{\ln\frac{x}{y}\pm(\sigma_1-\sigma_2)^2\cdot\frac{T}{2}}{(\sigma_1-\sigma_2)\sqrt{T}}$, *c is the unique solution to (4.61) if* $-q \leqslant 1$, $1 < c_1 \leqslant c_2$ *are two solutions to (4.61) in case* $-q > 1$.

Note that the results in (4.63) contain the Margrabe formula (see (2.53)) used to calculate the value $E^*(S_T^1 - S_T^2)^+$.

Remark 4.4. A constant c (or constants c_1 and c_2) can be found, as before, by fixing a probability ϵ of a failure to hedge the payoff H. Assuming $0 < \sigma_1 - \sigma_2 < \min(\sigma_1, \sigma_2)$, and $\frac{\mu_1 - r}{\sigma_1} = \frac{\sigma_1 \sigma_2}{\sigma_2 + (\sigma_1 - \sigma_2)^2}$, we find that $-q = \frac{\sigma_2 - (\sigma_2 - \sigma_1)\sigma_2}{\sigma_2 + (\sigma_1 - \sigma_2)^2} < 1$. Then equation (4.61) has a unique solution c, and A^* can be approximated by $\left\{\frac{S_T^1}{S_T^2} \leq c\right\}$ where $P\left(\frac{S_T^1}{S_T^2} \leq c\right) = 1 - \epsilon$. Using the log-normality of the ratio $\frac{S_T^1}{S_T^2}$, we arrive at the following formula

$$c = \exp\left\{\sqrt{T}(\sigma_1 - \sigma_2)\Phi^{-1}(1 - \epsilon) + \left(\mu_1 - \mu_2 - \frac{\sigma_1^2 - \sigma_2^2}{2}\right)T\right\}.$$

If the characteristic equation (4.61) has two solutions c_1 and c_2 (the case $q < -1$), they can be found from $P\left(\left\{\frac{S_T^1}{S_T^2} \leq c_1\right\} \cup \left\{\frac{S_T^1}{S_T^2} \geq c_2\right\}\right) = 1 - \epsilon$ in a similar way.

Theorem 4.5. *Consider a financial market with two risky assets following Model 3 and an equity-linked life insurance contract with payoff* $(S_T^1 - S_T^2)^+ \mathbb{1}_{\{T(x) > T\}}$ *linked to the evolution of the risky assets. We assume that the conditions (4.51) are fulfilled. Then, it is possible to balance the survival probability of an insured and the financial risk associated with the contract as follows:*

$$_Tp_x = 1 - \frac{\sum_{n=0}^{\infty} p_{n,T}^* \left(v_{n,T}^1 \Phi(b_+(v_{n,T}^1, d_n v_{n,T}^2, T)) - v_{n,T}^2 \Phi(b_-(v_{n,T}^1, d_n v_{n,T}^2, T))\right)}{\sum_{n=0}^{\infty} p_{n,T}^* \left(v_{n,T}^1 \Phi(b_+(v_{n,T}^1, v_{n,T}^2, T)) - v_{n,T}^2 \Phi(b_-(v_{n,T}^1, v_{n,T}^2, T))\right)},$$

(4.64)

where $v_{n,T}^i = (1 - \nu_i)^n e^{\nu_i}$, $i = 1, 2$, $p_{n,T}^* = \exp\{-\lambda^* T\} \frac{(\lambda^* T)^n}{n!}$, λ^* *is found from (4.52). Here the characteristic equation (4.61) on a set* $\{\Pi_T = n\}$ *admits only one solution* d_n *provided that* $\frac{(\mu_1 - r)\nu_1 - (\mu_2 - r)\nu_2}{\sigma_2 \nu_1 - \sigma_1 \nu_2} = \frac{\sigma_1 \sigma_2}{\sigma_2 + (\sigma_1 - \sigma_2)^2}$.

Note that the expression (4.64) represents the Margrabe formula weighted by the Poisson distribution $p_{n,T}^* = \exp\{-\lambda^* T\} \frac{(\lambda^* T)^n}{n!}$.

Remark 4.5. The constants d_n can be found by fixing the probability ϵ of a failure to hedge the payoff H on each set $\{\Pi_T = n\}$ or $P\left(\frac{S_T^1}{S_T^2} \leq d_n | \Pi_T = n\right) = 1 - \epsilon$. Using the log-normality of this conditional distribution, we find that

$$d_n = \left(\frac{1-\nu_1}{1-\nu_2}\right)^n \exp\left\{(\sigma_1 - \sigma_2)\Phi^{-1}(1-\epsilon)\sqrt{T} + \left(\mu_1 - \mu_2 - \frac{\sigma_1^2 - \sigma_2^2}{2}\right)T\right\}.$$

Here we emphasize again how to diversify the mortality risk. Mortality at a rate lower than expected would create unfunded claims exposing an insurance company to a mortality frequency risk. However, insurance companies are used to dealing with this type of risk. A large enough portfolio of life insurance contracts will result in more predictable mortality exposure and reduced risk.

Quantile hedging and risk management of contracts

We have dealt before with a single contract and found that the price of an equity-linked life insurance contract with a maturity guarantee H will be equal to $_Tp_x E^* H e^{-rT}$ which is less than the Black–Scholes or the Margrabe price $E^* H e^{-rT}$ for a similar contract with a fixed or flexible guarantee. The insurance company is able to take advantage of the diversification of the mortality risk that enables it to further reduce the price of the contract.

Here, we pool together the homogeneous clients of the same age, life expectancy and the same investment preferences and consider cumulative claims $l_{x+T} \cdot H$, where l_{x+T} is the number of insured alive at time T from the group of size l_x. If $H_{0,x}$ is an initial capital of a quantile hedge $\pi = \pi_\epsilon$ with the risk level ϵ, then

$$P(X_T^\pi \geqslant H) = 1 - \epsilon.$$

The maximal successful hedging set is invariant with respect to a multiplication by a positive constant b. Hence, $\pi = \pi_\epsilon$ represents a quantile hedge for the claim $b \cdot H$ with the same risk level and the initial price $b \cdot H_{0,\epsilon}$. Take $b = \frac{n_\delta}{l_x}$, where n_δ is defined by

$$\tilde{P}(n_\delta \geqslant l_{x+T}) = 1 - \delta.$$

The parameter $\delta \in (0,1)$ characterizes the level of a cumulative mortality risk or, in other words, the probability that the number of clients alive at maturity will be greater than expected based on the life expectancy of homogeneous clients. Having a frequency distribution, this probability could be calculated with the help of a binomial distribution with parameters $_Tp_x$ and l_x. Due to the independence of the insurance and the financial risks, we have

$$P \times \tilde{P}(l_x X_T^\pi \geqslant l_{x+T} H) \geqslant P\left(X_T^\pi \geqslant \frac{n_\delta}{l_x} H\right) \tilde{P}(n_\delta \geqslant l_{x+T})$$
$$\geqslant (1-\epsilon)(1-\delta) \geqslant 1 - (\epsilon + \delta).$$

So, using the strategy $\pi = \pi_\epsilon$ and the initial price $H_{\epsilon,\delta} = \frac{n_\delta}{l_x} H_{0,\epsilon}$, the insurance company can hedge the cumulative claim with the probability at least $1 - (\epsilon + \delta)$ combining both financial and insurance risks. In other words, the price of a single contract in the group can be calculated as $\frac{n_\delta}{l_x} {}_T p_x E^* H e^{-rT}$, and this reduced price will be associated with the cumulative risk $(\epsilon + \delta)$.

Now we are ready to formulate a scheme for risk-management of equity-linked life insurance contracts using quantile hedging methodology.

Step 1. Fix the level of a financial risk ϵ or the probability that the insurance company will fail to hedge successfully the claim H with a maturity guarantee.

Step 2. Find a survival probability $_Tp_x$ of a client whose mortality risk would offset a financial risk chosen using formulae (4.62)-(4.64). The obtained survival probability will give a quantile price of the contract $_Tp_x E^* H e^{-rT}$.

Step 3. Fix the level of an insurance risk δ or the probability that the number of clients alive at maturity will be greater than expected.

Step 4. Using a binomial distribution (or other appropriate frequency distribution) with parameters l_x and $_Tp_x$, where l_x is the number of homogeneous clients of age x with the same investment preferences, calculate the number n_δ of clients that would expose the insurance company to a cumulative mortality risk α at maturity.

Step 5. Calculate the price of the contract $\frac{n_\delta}{l_x} {_Tp_x} E^* H e^{-rT}$ that reflects both financial and insurance risks.

Step 6. Repeat the same steps for all possible combinations of ϵ and δ. Design a grid for financial and insurance risks reflecting the corresponding values of equity-linked insurance contracts as a function of the values of risks.

The risk-management scheme presented in Steps 1–6 is displayed in Figure 4.1.

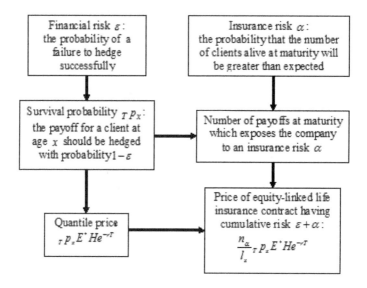

FIGURE 4.1: Risk-management scheme.

Let's realize the scheme in Figure 4.1 using numerical analysis.

Example 4.2. Estimation of the parameters of a diffusion and, especially, a jump-diffusion model represents a challenge in itself. Moreover, every model is very sensitive to its assumptions. Since these issues lie beyond the scope of our book, we use simple approaches proposed by others. We estimate the parameters of Models 1 and 2 using standard statistical methods and slightly modify the known approach to get estimates for the parameters of Model 3. In particular, there is one Poisson process in Model 3 that determines jumps in the prices for two assets, so we have to estimate instants of jumps from the evolution of both processes. The following idea is used: a Poisson process has jumped within $(t_{i-1}, t_i]$ if and only if the increment of the return Δr_{t_i} is too large in absolute value. As a result, we indicate that a jump occurs when and if $\left\{(\Delta r_{t_i}^1)^2 > h^1\right\}$ and $\left\{(\Delta r_{t_i}^2)^2 > h^2\right\}$ simultaneously, where $r_{t_i}^1$ and $r_{t_i}^2$ are monthly returns on risky-assets, h^1, h^2 are some functions defined as follows

$$h^i = 8M^2 \beta c \ln \frac{1}{c}, \quad |\mu_i| \leq M, \quad |\sigma_i| \leq M, \quad c \to 0, \quad \beta \in (1, 2].$$

We consider the financial indices Russell 2000 and the S&P 500 as risky assets. It is supposed that the first index is more risky than the second one as it consists of small stocks. Therefore, we consider the S&P 500 as a flexible guarantee. For the contracts with a fixed guarantee, the S&P 500 is used as a risky asset. We work with monthly observations over 25 years from January 1979 to December 2004 to estimate the parameters of Models 1, 2 and 3. The following estimates were obtained.

Model 1: $\mu = 0.14, \quad \sigma = 0.15.$

Model 2: $\mu_1 = 0.15, \quad \sigma_1 = 0.20;$
$\mu_2 = 0.14, \quad \sigma_2 = 0.15.$

Model 3: $\lambda = 0.07;$
$\mu_1 = 0.11, \quad \sigma_1 = 0.20, \quad \nu_1 = 0.04;$
$\mu_2 = 0.10, \quad \sigma_2 = 0.16, \quad \nu_2 = 0.04.$

For the contracts with a fixed guarantee, we use $k = 1$ as the coefficient of a fixed guarantee. The initial investment is $S_0 = 100$, terms of the contracts are $T = 5, 10, 15, 20, 25$ years and the risk-free rate is $r = 6\%$.

The formulae from Theorems 4.3–4.5 were used to calculate the survival probabilities of the clients for different levels of financial risk ϵ. The corresponding ages of the clientele are found from the mortality table UP94@2015 which represents best estimates for mortality projected to the year 2015. The results are displayed in Figure 4.2.

It shows that the methodology proposed in this works better for riskier contracts that are contracts with flexible guarantees and shorter duration contracts with fixed guarantees. It is clearly shown on the first graph of Figure 4.2 for the contracts with the fixed guarantee. Here, for durations longer than 15 years there is no financial risk to the insurance company. At the same time the contracts with a flexible guarantee of all durations expose the insurer to a financial risk. As a result, we further consider contracts with a fixed guarantee of shorter durations only. The results for Models 2 and 3 seem to be different which could be explained by the necessity to use more refined and consistent methods for parameter estimation which, however, lies beyond our scope here.

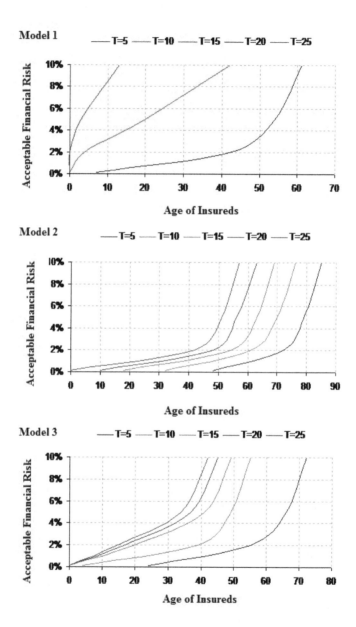

FIGURE 4.2: Offsetting financial and mortality risks.

Model 1

Age of Clients	$T=5$	$T=10$	$T=15$
30	1.3%	6.6%	16.3%
35	1.5%	7.6%	18.6%
40	1.9%	9.1%	22.6%
45	2.4%	11.7%	24.6%
50	3.4%	16.7%	39.6%
55	5.5%	24.6%	51.7%

Model 2

Age of Clients	$T=5$	$T=10$	$T=15$	$T=20$	$T=25$
30	0.04%	0.09%	0.17%	0.28%	0.48%
35	0.04%	0.11%	0.22%	0.4%	0.77%
40	0.06%	0.15%	0.34%	0.7%	1.4%
45	0.08%	0.24%	0.58%	1.3%	2.7%
50	0.13%	0.45%	1.13%	2.5%	4.6%
55	0.27%	0.90%	2.20%	4.4%	8%

Model 3

Age of Clients	$T=5$	$T=10$	$T=15$	$T=20$	$T=25$
30	0.14%	1.2%	3.0%	3.3%	3.9%
35	0.16%	1.5%	3.4%	4.3%	5.6%
40	0.22%	2.3%	4.6%	6.3%	8.7%
45	0.30%	3.1%	6.9%	9.8%	13.6%
50	0.50%	5.5%	11.0%	15.3%	19.8%
55	1.00%	10%	17.4%	22.0%	28.5%

TABLE 4.3: Acceptable financial risk offsetting mortality risks.

Model 1

Age of Clients	$T=5$	$T=10$	$T=15$
30	28.45	46.05	59.05
35	28.44	45.99	58.88
40	28.41	45.87	58.53
45	28.37	45.64	57.74
50	28.27	45.10	54.96
55	28.03	43.52	50.49
Black–Scholes price	28.82	46.81	60.28

Model 2

Age of Clients	$T=5$	$T=10$	$T=15$	$T=20$	$T=25$
30	4.40	6.20	7.56	8.67	9.60
35	4.40	6.20	7.53	8.61	9.36
40	4.39	6.18	7.49	8.49	8.98
45	4.39	6.14	7.39	8.07	8.12
50	4.37	6.07	7.03	7.31	6.94
55	4.34	5.86	6.46	6.27	5.26
Margrabe price	4.46	6.30	7.71	8.90	9.95

Model 3

Age of Clients	$T=5$	$T=10$	$T=15$	$T=20$	$T=25$
30	4.18	5.89	7.18	8.23	9.12
35	4.18	5.88	7.15	8.18	8.89
40	4.17	5.86	7.11	8.06	8.53
45	4.17	5.83	7.02	7.66	7.71
50	4.15	5.77	6.68	6.94	6.59
55	4.12	5.56	6.14	5.95	5.00
Margrabe price	4.23	5.98	7.32	8.45	9.45

TABLE 4.4: Prices of equity-linked life insurance contracts with $\delta = 2.5\%$.

Note also that whenever the risk that the insurance company will fail to hedge successful hedging increases, the recommended ages of the clients rise as well. As a result, the company diminishes the insurance component of the risk by attracting older and, therefore, safer clientele to compensate for the increasing financial risk. We also observe that with longer contract maturities, the company can widen its audience to younger clients because a mortality risk, which is a survival probability in our case, is decreasing over time. Also, contracts with a flexible guarantee have greater exposure to a financial risk than similar contracts with a fixed guarantee. As a result, the mortality risk for such contracts should be relatively lower.

The next step is to construct a grid that enables us to identify the acceptable level of a financial risk for insured of any age. We restrict our attention to a group of clients of ages 30, 35, 40, 45, 50, and 55 years. The results are presented in Table 4.3. The financial risk found reflects the acceptable probability of a failure to hedge the payoff for clients at these ages.

Prices of the contracts for the same group of clients are given in Table 4.4. Note that the price of a contract is a function of the financial and an insurance risks associated with this contract. The level of insurance risk is chosen to be $\delta = 2.5\%$. In the last rows, the Black–Scholes and Margrabe prices are compared with the reduced prices of equity-linked contracts. The reduction in the prices was possible for two reasons: we took into account the mortality risk of an individual client (the probability that the client would not survive to maturity and, therefore, no payment at maturity would be made) and the possibility to diversify the cumulative mortality risk by pooling homogeneous clients.

Proof of Theorem 4.5 To construct a successful hedging set we represent $Y_T = \frac{S_T^1}{S_T^2}$ as follows

$$Y_T = \left(\frac{1-\nu_1}{1-\nu_2}\right)^{\Pi_T} \exp\left\{(\sigma_1 - \sigma_2)W_T + \left(\mu_1 - \mu_2 - \frac{\sigma_1^2 - \sigma_2^2}{2}\right)T\right\}$$

$$= \left(\frac{1-\nu_1}{1-\nu_2}\right)^{\Pi_T} \exp\left\{(\sigma_1 - \sigma_2)W_T^* + \left((\nu_1 - \nu_2)\lambda^* - \frac{\sigma_1^2 - \sigma_2^2}{2}\right)T\right\},$$

where $W_t^* = W_t - \alpha^* t$ is a Wiener process with respect to P*.

Using these formulae and the representation $W_T = (1 + (\sigma_2 - \sigma_1))W_T - (\sigma_2 - \sigma_1)W_T$, we find that

$$Z_T^* S_T^2 = (S_T^1)^{\frac{1+(\sigma_2-\sigma_1)}{\sigma_1}\alpha^*} (S_T^2)^{1-\frac{\sigma_2-\sigma_1}{\sigma_1}\alpha^*} \cdot \text{const}^{\Pi_T} \cdot \text{const}.$$

Now, if the following condition

$$q = \frac{1+(\sigma_2-\sigma_1)}{\sigma_1}\alpha^* = \frac{\sigma_2-\sigma_1}{\sigma_2}\alpha^* - 1$$

is fulfilled, then $Z_T^* S_T^2 = Y_T^q \cdot \text{const}^{\Pi_T} \cdot \text{const}$.

In this case $\alpha^* = \frac{-\sigma_1\sigma_2}{\sigma_2+(\sigma_1-\sigma_2)^2}$ and the following condition should be satisfied

$$\frac{(\mu_1 - r)\nu_1 - (\mu_2 - r)\nu_2}{\sigma_2\nu_1 - \sigma_1\nu_2} = \frac{\sigma_1\sigma_2}{\sigma_2 + (\sigma_1 - \sigma_2)^2}.$$

Further, we consider the successful hedging set on $\{\Pi_T = n\}$, $n = 0, 1, \ldots$ and arrive at the characteristic equation (4.61)

$$y^{-\frac{1+(\sigma_2-\sigma_1)}{\sigma_1}\alpha^*} = \text{const}\,(y-1)^+.$$

Under the above conditions this equation admits only one solution $d_n \geq 1$, $n = 0, 1, \ldots$ and the successful hedging set is equal to $\{Y_T \leq d_n\}$, $n = 0, 1, \ldots$ on the set $\{\Pi_T = n\}$, $n = 0, 1, \ldots$.

Now we could use (4.60) to derive a concrete expression for a survival probability in terms of the parameters of the Model 3. First of all, we note that the denominator of (4.60) is given by the Margrabe formula weighted by the Poisson distribution (see Lemma 4.1). Further, the nominator of (4.60) can be represented as

$$\mathrm{E}^*\left[(S_T^1 - S_T^2)^+ \mathbb{1}_{\{Y_T \leq d_n\}} | \Pi_T = n\right] = \mathrm{E}^*\left[(S_T^1 - S_T^2)^+ | \Pi_T = n\right]$$
$$- \mathrm{E}^*\left[(S_T^1 - S_T^2) | \Pi_T = n\right] - \mathrm{E}^*\left[S_T^1 \mathbb{1}_{\{Y_T \leq d_n\}} | \Pi_T = n\right]$$
$$- \mathrm{E}^*\left[S_T^2 \mathbb{1}_{\{Y_T \leq d_n\}} | \Pi_T = n\right].$$

The first term of the right-hand side of the above equality coincides with the denominator of (4.60). The second one is easily calculated based on a martingale property of prices.

The other two terms are calculated with the help of Lemma 2.3 due to a lognormality of prices S_T^i, $i = 1, 2$ and their ratio Y_T conditioned on Π_T. After all, we directly arrive at the expression for a survival probability given in (4.64).

Chapter 5

CVaR-Hedging: Theory and applications

5.1 CVaR-hedging methodology and general theoretical facts 107
5.2 CVaR-hedging in the Black–Scholes model with applications to equity-linked life insurance 115
5.3 CVaR-hedging in the regime-switching telegraph market model 125

5.1 CVaR-hedging methodology and general theoretical facts

We approach the problem of partial hedging by minimizing conditional value-at-risk (CVaR), a quantile downside risk measure. Consider a probability space $(\Omega, \mathcal{F}, \mathrm{P})$ and a choice-dependent $\mathcal{F}$-measurable random variable $L(x)$ characterizing the loss, with strategy vector $x \in X$ and strategy constraints X. We assume that $\mathrm{E}_\mathrm{P}[|L(x)|] < \infty$ for all $x \in X$, where E_P means expectation w.r. to P.

Let $L_{(\alpha)}(x)$ and $L^{(\alpha)}(x)$ be lower and upper α-quantiles of $L(x)$:

$$L_{(\alpha)}(x) = \inf\{t \in \mathbb{R} : \mathrm{P}[L(x) \leq t] \geq \alpha\},$$

$$L^{(\alpha)}(x) = \inf\{t \in \mathbb{R} : \mathrm{P}[L(x) \leq t] > \alpha\}.$$

For a given strategy x and a fixed confidence level $\alpha \in (0,1)$, which in application would be a value fairly close to 1, value-at-risk (VaR) is defined as an upper α-quantile of the corresponding loss function,

$$\mathrm{VaR}^\alpha(L) = L^{(\alpha)}.$$

Conditional value-at-risk (CVaR), standing for conditional expected value of loss exceeding value-at-risk, is defined as

$$\mathrm{CVaR}^\alpha(L) = \inf\left\{z + \frac{1}{1-\alpha}\mathrm{E}_\mathrm{P}[(L(x)-z)^+] : z \in \mathbb{R}\right\}.$$

Although it may be not obvious from looking at the formal definition, CVaR is closely related to the notion of tail conditional expectation (TCE). Indeed, in a smooth case, when

$$P[L \geq L^{(\alpha)}, L \neq L(\alpha)] = 0,$$

conditional value-at-risk coincides with both upper and lower TCE:

$$\text{TCE}_\alpha(L) = \mathrm{E}[L|L \geq L_{(\alpha)}], \quad \text{TCE}^\alpha(L) = \mathrm{E}[L|L \geq L^{(\alpha)}].$$

Now, it turn out that it is possible to obtain values of both VaR and CVaR simultaneously by solving a certain one-dimensional convex optimization problem (see Theorem 5.1 of Rockafellar and Uryasev).

Let us define a special function $F_\alpha(x, z)$ to be used in the minimization theorem:

$$F_\alpha(x, z) = z + \frac{1}{1-\alpha}\mathrm{E}_P[(L(x) - z)^+]. \tag{5.1}$$

Theorem 5.1. *As a function of z, function $F_\alpha(x, z)$ defined by (5.1) takes on finite value is convex (hence continuous), and*

$$\mathrm{CVaR}_\alpha(x) = \min_{z \in \mathbb{R}} F_\alpha(x, z),$$

$$\mathrm{VaR}_\alpha(x) = \min\{y | y \in \mathrm{argmin}_{z \in \mathbb{R}} F_\alpha(x, z)\}.$$

In particular, one always has

$$\mathrm{VaR}_\alpha(x) \in \mathrm{argmin}_{z \in \mathbb{R}} F_\alpha(x, z),$$

$$\mathrm{CVaR}_\alpha(x) = F_\alpha(x, \mathrm{VaR}_\alpha(x)).$$

Remark 5.1. Theorem 5.1 sheds light on the question of why CVaR is a more stable performance criterion than VaR : it is well known in optimization theory that the optimal value generally admits a more robust behavior than the argminimum.

An important corollary is that the problem of CVaR minimization may be expressed as a problem of $F_\alpha(x, z)$ minimization.

Corollary 5.1. *Minimization of $\mathrm{CVaR}_\alpha(x)$ over the strategy set X is equivalent to minimization of $F_\alpha(x, z)$ over $X \times \mathbb{R}$:*

$$\min_{x \in X} \mathrm{CVaR}_\alpha(x) = \min_{x \in X} \min_{z \in \mathbb{R}} F_\alpha(x, z).$$

Let the discounted price process X_t be a semimartingale on a standard stochastic basis $(\Omega, \mathcal{F}, (\mathcal{F}_t)_{t \in [0,T]}, \mathrm{P})$ with $\mathcal{F}_0 = \{\emptyset, \Omega\}$ (see Chapter 1).

A self-financing strategy is defined by initial wealth $V_0 > 0$ and a predictable process ξ_t determining portfolio dynamics. For each strategy (V_0, ξ) the corresponding value process V_t is

$$V_t = V_0 + \int_0^t \xi_s dX_s, \quad t \in [0, T].$$

We shall call a strategy (V_0, ξ) admissible if it satisfies

$$V_t \geq 0, \text{ for all } t \in [0,T], \text{ P} - \text{a.s.,}$$

and we shall denote the set of all admissible self-financing strategies by $\mathcal{A}$.

Consider a discounted contingent claim whose payoff is an $\mathcal{F}_T$-measurable nonnegative random variable $H \in L^1(\text{P})$. In a complete market, there exists a unique equivalent martingale measure $\text{P}^* \sim \text{P}$, and construction of a perfect hedge is always possible. The perfect hedging strategy requires allocating the initial capital in the amount of

$$H_0 = \text{E}_{\text{P}^*}[H].$$

The first question is: if, for some reason, it is impossible to allocate the required amount of initial wealth H_0 for hedging, what is the best hedge that can be constructed using a smaller amount $\tilde{V}_0 < H_0$? Evidently, perfect hedging cannot be used in this case; instead, we have access to an infinite set of partial hedges, and to come to determination we need to fix an optimality criterion. As such, conditional value-at-risk (CVaR) risk measure shall be used.

We define the loss function from the viewpoint of a claim seller who hedges a short position in H with portfolio (V_0, ξ), thus the loss at time T equals the claim value less the terminal value of the hedging portfolio:

$$L(V_0, \xi) = H - V_T = H - V_0 - \int_0^T \xi_s dX_s. \tag{5.2}$$

Consider a fixed confidence level α and a strategy (V_0, ξ)

$$\text{CVaR}_\alpha(V_0, \xi) = \text{E}_\text{P}[L(V_0, \xi)|L(V_0, \xi) \geq \text{VaR}_\alpha(V_0, \xi)].$$

So, our first problem is to find an admissible strategy (V_0, ξ) which minimizes CVaR_α while not using more initial wealth than $\tilde{V}_0$.

Another question relates to the dual problem: what is the least amount of the initial capital we have to put up to keep CVaR of a given confidence level below a certain threshold. Again, it may be formulated as an optimization problem.

Both problems will be discussed in full below.

We suggest a method of solving the problem of CVaR minimization subject to a constraint:

$$\begin{cases} \text{CVaR}_\alpha(V_0, \xi) \to \min_{(V_0, \xi) \in \mathcal{A}}, \\ V_0 \leq \tilde{V}_0. \end{cases} \tag{5.3}$$

For simplicity of notation, denote by $\mathcal{A}_{\tilde{V}_0}$ the set of all admissible strategies satisfying the capital constraint:

$$\mathcal{A}_{\tilde{V}_0} = \left\{ (V_0, \xi) | (V_0, \xi) \in \mathcal{A}, V_0 \leq \tilde{V}_0 \right\}.$$

According to Corollary 5.1, problem (5.3) is equivalent to the following one:

$$F_\alpha((V_0,\xi),z) \to \min_{(V_0,\xi)\in\mathcal{A}_{\tilde{V}_0}} \min_{x\in\mathbb{R}}.$$

Recall that F_α is given by (5.1) and that the loss function for this problem is given by (5.2), so the original problem becomes

$$z + \frac{1}{1-\alpha} \cdot E_P[(H - V_T - z)^+] \to \min_{(V_0,\xi)\in\mathcal{A}_{\tilde{V}_0}} \min_{x\in\mathbb{R}}.$$

Let us introduce a function

$$c(z) = z + \frac{1}{1-\alpha} \cdot \min_{(V_0,\xi)\in\mathcal{A}_{\tilde{V}_0}} E_P[(H - V_T - z)^+], \qquad (5.4)$$

such that

$$\min_{(V_0,\xi)\in\mathcal{A}_{\tilde{V}_0}} \mathrm{CVaR}_\alpha(V_0,\xi) = \min_{x\in\mathbb{R}} c(z).$$

Assume that for each $z \in R$ the minimum in (5.4) is attained at $(\tilde{V}_0(z), \tilde{\xi}(z))$ and that $c(z)$ reaches its global minimum at point $\tilde{z}$:

$$\min_{(V_0,\xi)\in\mathcal{A}_{\tilde{V}_0}} E_P[(H - V_T - z)^+] = E_P[(H - \tilde{V}_T(z) - z)^+],$$

$$\min_{x\in\mathbb{R}} c(z) = c(\tilde{z}).$$

Then strategy $(\tilde{V}_0, \tilde{\xi}) = (\tilde{V}_0(\tilde{z}), \tilde{\xi}(\tilde{z}))$ is a solution for (5.3):

$$\min_{(V_0,\xi)\in\mathcal{A}_{\tilde{V}_0}} \mathrm{CVaR}_\alpha(V_0,\xi) = \mathrm{CVaR}_\alpha(\tilde{V}_0(\tilde{z}), \tilde{\xi}(\tilde{z})).$$

Definition (5.4) of function $c(z)$ contains expected value minimization. Deriving explicit expression for this function would provide the possibility to reduce the initial problem (5.3) to a problem of one-dimensional optimization; to do that we shall use some known results in the area of expected shortfall minimization.

For each z strategy $(\tilde{V}_0(z), \tilde{\xi}(z))$ is a solution for the following problem:

$$E_P[(H - V_T - z)^+] \to \min_{(V_0,\xi)\in\mathcal{A}_{\tilde{V}_0}}. \qquad (5.5)$$

Note that

$$(H - V_T - z)^+ \equiv ((H - z)^+ - V_T)^+.$$

It is not hard to see that $(H - z)^+$ is an $\mathcal{F}_T$-measurable, nonnegative random variable-therefore we can consider it as a contingent claim. Problem (5.5) then may be restated as

$$E_P\left[((H - z)^+ - V_T)^+\right] \to \min_{(V_0,\xi)\in\mathcal{A}_{\tilde{V}_0}} \qquad (5.6)$$

Problem (5.6) can be treated as a problem of expected shortfall minimization with respect to a contingent claim with payoff $(H - z)^+$ that depends on a real-valued parameter z. This kind of problem is well studied (see Theorem 5.2 below).

Theorem 5.2. (Föllmer and Leukert) *The optimal strategy $(\widetilde{V}_0(z), \widetilde{\xi}(z))$ for problem (5.6) is a perfect hedge for the contingent claim $\widetilde{H}(z) = (H - z)^+\widetilde{\varphi}(z)$:*

$$E_{P*}[\widetilde{H}(z)|\mathcal{F}_t] = \widetilde{V}_0(z) + \int_0^t \widetilde{\xi}_s(z) dX_s, \quad P-a.s., \ t \in [0,T], \quad (5.7)$$

where

$$\widetilde{\varphi}(z) = \mathbb{1}_{\{\frac{dP}{dP*} > \tilde{a}(z)\}} + \gamma(z)\mathbb{1}_{\{\frac{dP}{dP*} = \tilde{a}(z)\}}, \quad (5.8)$$

$$\tilde{a} = \inf\left\{a \geq 0 : E_{P*}\left[(H-z)^+ \cdot \mathbb{1}_{\{\frac{dP}{dP*} > \tilde{a}(z)\}}\right] \leq \widetilde{V}_0\right\}, \quad (5.9)$$

$$\gamma(z) = \frac{\widetilde{V}_0 - E_{P*}\left[(H-z)^+ \cdot \mathbb{1}_{\{\frac{dP}{dP*} > \tilde{a}(z)\}}\right]}{E_{P*}\left[(H-z)^+ \cdot \mathbb{1}_{\{\frac{dP}{dP*} = \tilde{a}(z)\}}\right]}. \quad (5.10)$$

Theorem 5.2 provides an explicit solution for (5.6) in terms of the Neyman–Pearson framework - that is, $\widetilde{\varphi}(z)$ may be interpreted as an optimal randomized test (see also Chapter 3). Note that $\gamma(z)$ equals zero if the distribution of the Radon–Nikodym derivative $\frac{dP}{dP*}$ is atomless.

Let us summarize the results in the following theorem.

Theorem 5.3. *The optimal strategy $(\widetilde{V}_0, \widetilde{\xi})$ for the problem of CVaR minimization (5.3) is a perfect hedge for the contingent claim $\widetilde{H}(\widetilde{z}) = (H - \widetilde{z})^+\widetilde{\varphi}(\widetilde{z})$, where $\widetilde{\varphi}(z)$ is defined by (5.8)-(5.10), $\widetilde{z}$ is a point of global minimum of function*

$$c(z) = \begin{cases} z + \frac{1}{1-\alpha}E_P\left[(H-z)^+(1-\widetilde{\varphi}(z))\right], & \text{for } z < z^*, \\ z, & \text{for } z \geq z^*, \end{cases} \quad (5.11)$$

on interval $z < z^*$, and z^* is the real root of equation

$$\widetilde{V}_0 = E_{P*}\left[(H-z^*)^+\right].$$

Besides, one always has

$$\text{CVaR}_\alpha(\widetilde{V}_0, \widetilde{\xi}) = c(\widetilde{z}), \quad (5.12)$$

$$\text{VaR}_\alpha(\widetilde{V}_0, \widetilde{\xi}) = \widetilde{z}. \quad (5.13)$$

We used the results of Theorem 5.2 to get rid of the minimum in the definition of $c(z)$. Note that problem (5.6) only makes sense when $\widetilde{V}_0 < E_{P*}[H(z)]$, (otherwise a perfect hedge for $H(z)$ may be used as an optimal strategy, providing zero expected shortfall). As a function of z, $E_{P*}[H(z)]$ is monotonous and non-increasing, and

$$E_{P*}[H(0)] = H_0 > \tilde{V}_0,$$

$$\lim_{z \to \infty} E_{P*}[H(z)] = 0,$$

so there exists $z^* > 0$ such that

$$\tilde{V}_0 \geq E_{P*}[H(z)], \quad \forall z \geq z^*. \tag{5.14}$$

Hence, when z is greater than z^*, the perfect hedge for $(H-z)^+$ can be used in problem (5.6) and this explains why $c(z) = z$ for $z \geq z^*$.

According to Theorem 5.1, the argminimum of $c(z)$ coincides with the value-at-risk of the CVaR-optimal hedge. Note that the loss function is always non-negative,

$$L(z) = H - \tilde{H}(z) = H - \tilde{\varphi}(z)(H-z)^+ \geq 0,$$

therefore the corresponding value-at-risk would also be nonnegative, so $\tilde{z} > 0$; besides, $c(z) = z$ for $z > z^*$ and $c(z)$ is increasing at $z = z^*$, so the global minimum of $c(z)$ coincides with its minimum on $(0, z^*)$.

Now we minimize the initial wealth over all admissible strategies (V_0, ξ) with conditional value-at-risk of a given confidence level not exceeding the predefined threshold $\tilde{C}$:

$$\begin{cases} V_0 \to \min_{(V_0, \xi) \in \mathcal{A}}, \\ \mathrm{CVaR}_\alpha(V_0, \xi) \leq \tilde{C}. \end{cases} \tag{5.15}$$

Let us rephrase the problem in terms of terminal capital $V_T = V_0 + \int_0^T \xi_s dX_s$ (we can always go back and derive the trading strategy explicitly by constructing a perfect hedge):

$$\begin{cases} E_{P*}[V_T] \to \min_{V_T \in \mathcal{F}_T}, \\ \mathrm{CVaR}_\alpha(V_T) \leq \tilde{C}. \end{cases} \tag{5.16}$$

Recall that

$$\mathrm{CVaR}_\alpha(V_0, \xi) = \min_{x \in \mathbb{R}} \left(z + \frac{1}{1-\alpha} E_P(H - V_T - z)^+ \right),$$

and consider a family of problems

$$\begin{cases} E_{P*}[V_T] \to \min_{V_T \in \mathcal{F}_T}, \\ E_P(H - V_T - z)^+ \leq (\tilde{C} - z)(1 - \alpha). \end{cases} \tag{5.17}$$

For consistency of notation, we provide the following lemma, which will be applied to problem (5.17).

Lemma 5.1. *Let $\tilde{x}$ be a solution of*

$$\begin{cases} f(x) \to \min_{x \in X}, \\ \min_{x \in \mathbb{R}} g(x, z) \leq c. \end{cases}$$

Then the following family of problems also admits solutions, denoted $\tilde{x}(z)$:

$$\begin{cases} f(x) \to \min_{x \in X}, \\ g(x, z) \leq c. \end{cases}$$

Besides, one always has

$$\tilde{x} = \tilde{x}(\tilde{z}),$$

where $\tilde{z}$ is a point of global minimum of $f(\tilde{x}(z))$:

$$\min_{x \in \mathbb{R}} f(\tilde{x}(z)) = f(\tilde{x}(\tilde{z})).$$

Proof. Indeed, for each $z \in \mathbb{R}$:

$$\bigcup_{x \in \mathbb{R}} \{x | g(x, z) \leq c\} = \left\{ x \middle| \min_{x \in \mathbb{R}} g(x, z) \leq c \right\},$$

and

$$\bigcup_{x \in \mathbb{R}} [\mathbb{X} \cap \{x | g(x, z) \leq c\}] = \mathbb{X} \cap \left\{ x \middle| \min_{x \in \mathbb{R}} g(x, z) \leq c \right\}.$$

Therefore,

$$\min_{x \in \mathbb{X} \cap \left\{ x | \min_{x \in \mathbb{R}} g(x,z) \leq c \right\}} f(x) = \min_{x \in \mathbb{R}} \left[\min_{x \in \mathbb{X} \cap \{x | g(x,z) \leq c\}} f(x) \right],$$

which proves the lemma. □

Denote the solution of (5.17) for each real z by $\tilde{V}_T(z)$, then, according to the lemma stated above, the solution for (5.16) may be expressed as

$$\tilde{V}_T = \tilde{V}_T(\tilde{z}), \tag{5.18}$$

where

$$\mathrm{E}_{\mathrm{P}*}[\tilde{V}_T(\tilde{z})] = \min_{x \in \mathbb{R}} \mathrm{E}_{\mathrm{P}*}[\tilde{V}_T(z)]. \tag{5.19}$$

We shall derive $\tilde{V}_T$ by solving (5.17). To start with, note that in the case $z > c$ the problem admits no solution since the left side is always nonnegative. In the case $z \leq c$ note that

$$(H - V_T - z)^+ = ((H - z)^+ - V_T)^+,$$

and, since $0 \leq V_T \leq (H-z)^+$, let

$$V_T = (H-z)^+(1-\varphi), \quad \varphi \in \mathcal{P}_{[0,1]},$$

where $\mathcal{P}_{[0,1]}$ is the class of $\mathcal{F}_T$-measurable random variables taking on values in $[0,1]$.

The initial problem can be then rewritten in terms of φ (its solution we will denote by $\widetilde{\varphi}(z)$):

$$\begin{cases} E_P\left[(H-z)^+\varphi\right] \leq (\widetilde{C}-1)(1-\alpha), \\ E_{P^*}\left[(H-z)^+\varphi\right] \to \max_{\varphi \in \mathcal{P}_{[0,1]}}. \end{cases}$$

This problem can be solved by applying the Neyman–Pearson lemma, and it only makes sense as long as $E_P[(H-z)^+] > (\widetilde{C}-z)(1-\alpha)$, otherwise we can set $\widetilde{\varphi}(z) \equiv 1$ and $\widetilde{V}_T(z) \equiv 0$.

Lemma 5.2. *The following condition:*

$$E_P\left[(H-z)^+\right] > (\widetilde{C}-z)(1-\alpha), \tag{5.20}$$

is satisfied for all $z \leq \widetilde{C}$ if and only if two inequalities hold true:

$$E_P[H] > \widetilde{C}(1-\alpha), \quad E_P[(H-\widetilde{C})^+] > 0. \tag{5.21}$$

Proof. Note that both right and left sides of (5.20) are monotonous non-increasing functions of z. In addition,

$$\frac{d}{dz}E_P[(H-z)^+] = -1, \quad \text{for } z < 0,$$

$$\frac{d}{dz}(\widetilde{C}-z)(1-\alpha) = -1+\alpha,$$

so it is necessary and sufficient that (5.20) holds true at points $z=0$ and $z=\widetilde{C}$ only, which implies (5.21) and thus proves the lemma. □

Remark 5.2. Lemma 5.2 provides an easy way to check whether (5.20) is satisfied for all $z \leq \widetilde{C}$ or not: if there exists such $z = z^*$ that it does not hold true, then $\widetilde{V}_T(z^*) \equiv 0$ and, according to (5.18) and (5.19), the solution for (5.16) would be also equal to zero, which can be interpreted as selecting a passive trading strategy. Indeed, if the first inequality in (5.21) is not satisfied, the target CVaR is too high compared to the expected payoff on the contingent claim, so there is no need to hedge at all; if the second inequality is not satisfied, the payoff is bounded from above by a constant less than $\widetilde{C}$, so CVaR can never reach its target value no matter what strategy is used.

Theorem 5.4. *The optimal strategy $(\widetilde{V}_0, \widetilde{\xi})$ for the problem of hedging costs minimization (5.15) is*

(a) A perfect hedge for the contingent claim $(H-\tilde{z})^+(1-\tilde{\varphi}(\tilde{z}))$, if condition (5.21) holds true, where $\tilde{\varphi}(z)$ is defined by

$$\tilde{\varphi}(z) = \mathbb{1}_{\{\frac{dP^*}{dP} > \tilde{a}(z)\}} + \gamma(z)\mathbb{1}_{\{\frac{dP^*}{dP} = \tilde{a}(z)\}},$$

$$\tilde{a} = \inf\left\{a \geqslant 0 : E_P\left[(H-z)^+ \cdot \mathbb{1}_{\{\frac{dP^*}{dP} > a\}}\right] \leqslant (\tilde{C}-z)(1-\alpha)\right\},$$

$$\gamma(z) = \frac{(\tilde{C}-z)(1-\alpha) - E_P\left[(H-z)^+ \cdot \mathbb{1}_{\{\frac{dP^*}{dP} > \tilde{a}(z)\}}\right]}{E_P\left[(H-z)^+ \cdot \mathbb{1}_{\{\frac{dP^*}{dP} = \tilde{a}(z)\}}\right]},$$

and $\tilde{z}$ is a point of minimum of function

$$d(z) = E_{P^*}[(H-z)^+(1-\tilde{\varphi}(z))]$$

on interval $-\infty < z \leqslant \tilde{C}$;

(b) A passive trading strategy, if condition (5.21) is not satisfied.

5.2 CVaR-hedging in the Black–Scholes model with applications to equity-linked life insurance

In the framework of the standard Black–Scholes model price of the underlying S_t and bond price B_t follow

$$\begin{cases} B_t = e^{rt}, \\ S_t = S_0 \exp(\sigma W_t + \mu t), \end{cases}$$

where r is the riskless interest rate, $\sigma > 0$ is the constant volatility, μ is the constant drift and W is a Wiener process under P.

We assume that there are no transaction costs and both instruments are freely tradable.

The SDE for the discounted price process $X_t = B_t^{-1} S_t$ is then given by

$$\begin{cases} dX_t = X_t(\sigma dW_t + m dt), \\ X_0 = x_0, \end{cases}$$

where $m = \mu - r + \frac{1}{2}\sigma^2$.

The unique equivalent martingale measure P* may be derived with the help of the Girsanov theorem:

$$\frac{dP^*}{dP} = \exp\left(-\frac{m}{\sigma}W_T - \frac{1}{2}\left(\frac{m}{\sigma}\right)^2 T\right). \tag{5.22}$$

Note that
$$X_T = x_0 \exp\left(\sigma W_T + \left(m - \frac{1}{2}\sigma^2\right)T\right),$$
so (5.22) may be rewritten as
$$\frac{dP^*}{dP} = \text{const} \cdot X_T^{-\frac{m}{\sigma^2}}. \tag{5.23}$$

The contingent claim of interest is a plain vanilla call option with payoff $(S_T - K)^+$. The discounted claim is also a call option with respect to X_T, with the strike price of Ke^{-rT}:
$$H = (X_t - Ke^{-rT})^+.$$

The initial capital H_0 required for a perfect hedge is
$$H_0 = E_{P^*}[H] = x_0 \Phi_+(Ke^{-rT}) - Ke^{-rT}\Phi_-(Ke^{-rT}),$$
where
$$\Phi_\pm(K) = \Phi\left(\frac{\ln x_0 - \ln K}{\sigma\sqrt{T}} \pm \frac{1}{2}\sigma\sqrt{T}\right), \tag{5.24}$$
and $\Phi(\cdot)$ is a cumulative distribution function for standard normal distribution.

In case the initial wealth is bounded above by $\tilde{V}_0 < H_0$, we cannot construct a perfect hedge for the call option; instead, we shall minimize CVaR over all admissible strategies with the initial wealth not exceeding $\tilde{V}_0$. The results of Paragraph 5.1 shall be used to derive the explicit solution.

As stated in Theorem 5.3, the original problem may be reduced to a problem of minimizing an auxiliary function $c(z)$ on interval $(0, z^*)$, where
$$c(z) = \begin{cases} z + \frac{1}{1-\alpha}E_P[(H - z)^+ \tilde{\varphi}(z)], & \text{for } z < z^*, \\ z, & \text{for } z \geqslant z^*, \end{cases}$$
$\tilde{\varphi}(z)$ is defined by (5.8)-(5.10) and z^* is real root of
$$\tilde{V}_0 = E_{P^*}[(H - z^*)^+]. \tag{5.25}$$

Since we consider $z > 0$ only,
$$(H - z)^+ = ((X_T - Ke^{-rT})^+ - z)^+ = (X_T - (Ke^{-rT} + z))^+.$$

For simplicity of notation, denote
$$H(z) = (X_T - K(z))^+,$$
$$K(z) = Ke^{-rT} + z,$$
$$\tilde{\Phi}_\pm(x) = \Phi_\pm(xe^{-mT}),$$
$$\Lambda_\pm = \Phi_\pm(x) - \Phi_\pm(y),$$
$$\tilde{\Lambda}_\pm = \tilde{\Phi}_\pm(x) - \tilde{\Phi}_\pm(y).$$

It is clear that $H(z)$ is also a call option, with the strike price of $K(z)$, so we can apply the Black–Scholes formula to (5.25):

$$\widetilde{V}_0 = x_0 \Phi_+(K(z^*)) - K(z^*)\Phi_-(K(z^*)). \tag{5.26}$$

Further on, we shall refer to z^* as the solution for (5.26).

We shall consider two cases.

(a) $\mu + \frac{1}{2}\sigma^2 > r$ ($m > 0$)

The set $\left\{\frac{dP}{dP^*} > a\right\}$ takes the form

$$\left\{\frac{dP}{dP^*} > a\right\} = \left\{X_T^{\frac{m}{\sigma^2}} > \widetilde{b}\right\} = \{X_T > b\},$$

and, moreover,

$$P\left(\frac{dP}{dP^*} = a\right) = P^*\left(\frac{dP}{dP^*} = a\right) = 0.$$

Applying this to (5.8)–(5.10), we get

$$\widetilde{\varphi}(z) = \mathbb{1}_{\{X_T > \widetilde{b}(x)\}},$$

$$\widetilde{b}(z) = \inf\left\{b \geq 0 : E_{P^*}[H(x) \cdot \mathbb{1}_{\{X_T > b\}}] \leq \widetilde{V}_0\right\},$$

$$\gamma(z) = 0.$$

Note that in our case the infimum is always attained since we deal with atomless measures. The expectation in the expression for $\widetilde{b}(z)$ may be rewritten as

$$E_{P^*}[H(z) \cdot \mathbb{1}_{\{X_T > b\}}] = \begin{cases} E_{P^*}[H(z)], & \text{for } b < K(z), \\ x_0\Phi_+(b) - K(z)\Phi_-(b), & \text{for } b \geq K(z). \end{cases}$$

Since we consider $z < z^*$, (5.14) applies:

$$E_{P^*}[H(z)] > \widetilde{V}_0,$$

therefore the minimum is not attained on the set $b < K(z)$, and hence $\widetilde{b}(z)$ is a solution for the following system:

$$\begin{cases} x_0\Phi_+(b) - K(z)\Phi_-(b) = \widetilde{V}_0, \\ b \geq K(z). \end{cases} \tag{5.27}$$

Note that the constraint in (5.27) is essential since the equation may have more than one real root; it is straightforward to show that for all $0 \leq z < z^*$ this system yields a single root $\widetilde{b}(z)$.

Now we are able to write down the function $c(z)$:

$$c(z) = z + \frac{1}{1-\alpha} \cdot E_P\left[\mathbb{1}_{\{X_T \leq \widetilde{b}(x)\}} \cdot H(z)\right],$$

or, evaluating the expectation

$$c(z) = z + \frac{1}{1-\alpha}\left(x_0 e^{mT}\tilde{\Lambda}_+(K(z),\tilde{b}(z)) - K(z)\tilde{\Lambda}_-(K(z),\tilde{b}(z))\right),$$

where $\tilde{b}(z)$ is a solution of (5.26).

According to Theorem 5.3, the optimal strategy $(\tilde{V}_0, \tilde{\xi})$ is then a perfect hedge for the contingent claim

$$\tilde{H}(\tilde{z}) = (H - \tilde{z})^+ \mathbb{1}_{\{X_T > \tilde{b}(\tilde{z})\}},$$

where $\tilde{z}$ is a point of a minimum of $c(z)$ on interval $(0, z^*)$.

(b) $\mu + \frac{1}{2}\sigma^2 < r$ $(m < 0)$
In this case the set $\{\frac{dP}{dP*} > a\}$ is

$$\left\{\frac{dP}{dP*} > a\right\} = \left\{X_T^{\frac{m}{\sigma^2}} > \tilde{b}\right\} = \{X_T < b\},$$

and therefore

$$\tilde{\varphi}(z) = \mathbb{1}_{\{X_T < \tilde{b}(x)\}},$$
$$\tilde{b}(z) = \sup\left\{b \geqslant 0 : \mathrm{E}_{\mathrm{P}*}[H(x) \cdot \mathbb{1}_{\{X_T < b\}}] \leqslant \tilde{V}_0\right\},$$
$$\gamma(z) = 0.$$

Denote
$$\beta(b, z) = x_0\Lambda_+(K(z), b) - K(z)\Lambda_-(K(z), b),$$
then
$$\mathrm{E}_{\mathrm{P}*}[H(z) \cdot \mathbb{1}_{\{X_T < b\}}] = \begin{cases} 0, & \text{for } b < K(z), \\ \beta(b, z), & \text{for } b \geqslant K(z). \end{cases}$$

Same as we did above, recall that $\mathrm{E}_{\mathrm{P}*}[H(z)] > \tilde{V}_0$ for $z < z^*$, and consider properties of $\beta(b, z)$:

$$\beta(K(z), z) = 0, \quad \beta(+\infty, z) = \mathrm{E}_{\mathrm{P}*}[H(z)], \quad \frac{\partial}{\partial b}\beta(b, x) \geqslant 0.$$

So, it is clear that the supremum is attained on the set $b \geqslant K(z)$ and hence $\tilde{b}(z)$ is a solution (which exists and is unique) for the following system:
$$\begin{cases} x_0\Lambda_+(K(z), b) - K(z)\Lambda_-(K(z), b) = \tilde{V}_0, \\ b \geqslant K(z). \end{cases} \quad (5.28)$$

Function $c(z)$ then takes from

$$c(z) = z + \frac{1}{1-\alpha} \cdot \mathrm{E}_{\mathrm{P}}\left[\mathbb{1}_{\{X_T \geqslant \tilde{b}(x)\}} \cdot H(z)\right],$$

or
$$c(z) = z + \frac{1}{1-\alpha}\left(x_0 e^{mT}\widetilde{\Phi}_+(\widetilde{b}(z)) - K(z)\widetilde{\Phi}_-(\widetilde{b}(z))\right),$$

where $\widetilde{b}(z)$ is a solution of (5.28) and $\Phi_\pm(\cdot)$ is defined by (5.24), and the optimal strategy $(\widetilde{V}_0, \widetilde{\xi})$ is a perfect hedge for the contingent claim

$$\widetilde{H}(\widetilde{z}) = (H - \widetilde{z})^+ \mathbb{1}_{\{X_T < \widetilde{b}(\widetilde{z})\}},$$

where $\widetilde{z}$ is a point of the minimum of $c(z)$ on interval $(0, z^*)$.

Example 5.1. To illustrate the method numerically, consider a financial market that evolves in accordance with the Black–Scholes model with parameters $\sigma = 0.3$, $\mu = 0.09$, $r = 0.05$ and a plain vanilla call option with the strike price of $K = 110$ and time to maturity $T = 0.25$. Let the initial price of the underlying be equal to $S_0 = 100$. In this setting, we are interested in hedging strategies that minimize $\text{CVaR}_{0.975}$ (conditional value-at-risk with a confidence level of 97.5%) for various amounts of the initial wealth. You can observe the results of computations in Figure 5.1.

FIGURE 5.1: Optimal CVaR.

Here we shall apply the results of Paragraph 5.1 to explicitly construct strategies minimizing the initial wealth with CVaR not exceeding target value $\widetilde{C}$.

According to Theorem 5.4, a passive trading strategy is optimal in hedging the cost minimization problem if at least one of inequalities (5.21) is not

satisfied. In the Black–Scholes setting, these inequalities take the form
$$x_0 e^{mT}\widetilde{\Phi}_+(K) - K\widetilde{\Phi}_-(K) - \widetilde{C}(1-\alpha) > 0,$$
$$x_0 e^{mT}\widetilde{\Phi}_+(K+\widetilde{C}) - (K+\widetilde{C})\widetilde{\Phi}_-(K+\widetilde{C}) > 0. \tag{5.29}$$

Further on we assume a non-trivial case, i.e., both inequalities in (5.29) are satisfied. Again, we consider two cases.

(a) $\mu + \frac{1}{2}\sigma^2 > r$ $(m > 0)$
In this case
$$\left\{\frac{d\mathrm{P}}{d\mathrm{P}^*} > a\right\} = \{X_T < b\}, \quad \mathrm{P}\left(\frac{d\mathrm{P}}{d\mathrm{P}^*} = a\right) = \mathrm{P}^*\left(\frac{d\mathrm{P}}{d\mathrm{P}^*} = a\right) = 0,$$
so we have
$$\widetilde{\varphi}(z) = \mathbb{1}_{\{X_T < \widetilde{b}(x)\}},$$
$$\widetilde{b}(z) = \sup\left\{b \geq 0 : \mathrm{E}_\mathrm{P}[H(x) \cdot \mathbb{1}_{\{X_T < b\}}] \leq (\widetilde{C} - z)(1-\alpha)\right\},$$
$$\gamma(z) = 0.$$

Denote
$$\delta(b, z) = x_0 e^{mT}\widetilde{\Lambda}_+(K(z), b) - K(z)\widetilde{\Lambda}_-(K(z), b),$$
then
$$\mathrm{E}_\mathrm{P}[H(z) \cdot \mathbb{1}_{\{X_T < b\}}] = \begin{cases} 0, & \text{for } b < K(z), \\ \delta(b, z), & \text{for } b \geq K(z), \end{cases}$$
$$\delta(K(z), z) = 0, \quad \delta(+\infty, z) = \mathrm{E}_\mathrm{P}[H(z)], \quad \frac{\partial}{\partial b}\delta(b, x) \geq 0.$$

Assuming that inequalities (5.29) hold true, the supremum is attained on the set $b \geq K(z)$, and hence $\widetilde{b}(z)$ is a unique solution of the system:
$$\begin{cases} x_0 e^{mT}\widetilde{\Lambda}_+(K(z), b) - K(z)\widetilde{\Lambda}_-(K(z), b) = (\widetilde{C} - z)(1-\alpha), \\ b \geq K(z). \end{cases} \tag{5.30}$$

The optimal strategy $(\widetilde{V}_0, \widetilde{\xi})$ would be a perfect hedge for the contingent claim $H(\widetilde{z})^+ \mathbb{1}_{\{X_T > \widetilde{b}(\widetilde{z})\}}$, where $\widetilde{b}(z)$ is a solution for (5.30) and $\widetilde{z}$ is a minimizer of function
$$d(z) = x_0 \Phi_+(\widetilde{b}(z)) - K(z)\Phi_-(\widetilde{b}(z))$$
on interval $z \in (-\infty, \widetilde{C})$.

(b) $\mu + \frac{1}{2}\sigma^2 < r$ $(m < 0)$
We have
$$\left\{\frac{d\mathrm{P}}{d\mathrm{P}^*} > a\right\} = \{X_T > b\}, \quad \mathrm{P}\left(\frac{d\mathrm{P}}{d\mathrm{P}^*} = a\right) = \mathrm{P}^*\left(\frac{d\mathrm{P}}{d\mathrm{P}^*} = a\right) = 0,$$

hence
$$\tilde{\varphi}(z) = \mathbb{1}_{\{X_T > \tilde{b}(x)\}},$$
$$\tilde{b}(z) = \sup\left\{b \geqslant 0 : \mathrm{E}_{\mathrm{P}}[H(x) \cdot \mathbb{1}_{\{X_T > b\}}] \leqslant (\tilde{C} - z)(1 - \alpha)\right\},$$
$$\gamma(z) = 0.$$

Denote
$$\zeta(b, z) = x_0 \tilde{\Phi}_+(b) - K(z)\tilde{\Phi}_-(b),$$
then
$$\mathrm{E}_{\mathrm{P}}[H(z) \cdot \mathbb{1}_{\{X_T > b\}}] = \begin{cases} \mathrm{E}_{\mathrm{P}}[H(z)], & \text{for } b < K(z), \\ \zeta(b, z), & \text{for } b \geqslant K(z), \end{cases}$$
$$\zeta(K(z), z) = \mathrm{E}_{\mathrm{P}}[H(z)], \quad \zeta(+\infty, z) = 0, \quad \frac{\partial}{\partial b}\zeta(b, x) \leqslant 0.$$

The supremum is attained on the set $b \geqslant K(z)$, and $\tilde{b}(z)$ is a unique solution of the system:
$$\begin{cases} x_0 \tilde{\Phi}_+(b) - K(z)\tilde{\Phi}_-(b) = (\tilde{C} - z)(1 - \alpha), \\ b \geqslant K(z). \end{cases} \quad (5.31)$$

The optimal strategy $(\tilde{V}_0, \tilde{\xi})$ would be a perfect hedge for the contingent claim $H(\tilde{z})^+ \mathbb{1}_{\{X_T < \tilde{b}(\tilde{z})\}}$, where $\tilde{b}(z)$ is a solution for (5.31) and $\tilde{z}$ is a point of the minimum of function
$$d(z) = x_0 \Lambda_+(K(z), \tilde{b}(z)) - K(z)\Lambda_-(K(z), \tilde{b}(z))$$
on interval $z \in (-\infty, \tilde{C})$.

Example 5.2. Now we apply the above results to the Black–Scholes model with the same parameters as in Example 5.1 ($\sigma = 0.3$, $\mu = 0.09$, $r = 0.05$, call option with the strike price of $K = 110$, time to maturity $T = 0.25$, initial price $S_0 = 100$). Figure 5.2 shows the minimum amount of the initial capital to be invested in the hedging strategy so that the resulting $\mathrm{CVaR}_{0.975}$ does not exceed a specified threshold.

Let us use the quantile methodology (see Chapter 2 and Chapter 4) to construct CVaR-optimal hedges of an embedded call option in an equity-linked life insurance contract.

In addition to the "financial" probability space $(\Omega, \mathcal{F}, \mathrm{P})$ introduced earlier, let us consider the "actuarial" probability space $(\tilde{\Omega}, \tilde{\mathcal{F}}, \tilde{\mathrm{P}})$. Let a random variable $T(x)$ denote the remaining lifetime of a person aged x, and let $_T p_x = \tilde{\mathrm{P}}(T(x) > T)$ be a survival probability for the next T years of the insured. We assume that $T(x)$ does not depend on the evolution of the financial market, so we can treat $(\Omega, \mathcal{F}, \mathrm{P})$ and $(\tilde{\Omega}, \tilde{\mathcal{F}}, \tilde{\mathrm{P}})$ as independent, providing our considerations on a product probability space, as we did before.

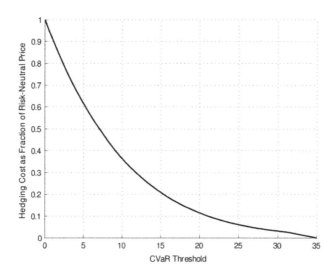

FIGURE 5.2: Hedging costs.

Under an equity-linked pure endowment contract, the insurance company is obliged to pay the benefit in the amount of $\overline{H}$ (an $\mathcal{F}_T$-measurable random variable) to the insured provided the insured is alive at time T. Essentially, the benefit is linked to the evolution of the financial market; hence an insurance contract of this kind poses two independent kinds of risk to the insurance company: mortality risk and market risk.

According to the option pricing theory, the optimal price is traditionally calculated as an expected present value of cash flows under the risk-neutral probability. However, the "insurance" part of the contract does not need to be risk adjusted since the mortality risk is essentially unsystematic. Denote the discounted benefit by $H = \overline{H} e^{-rT}$, then the price of the contract (Brennan–Schwartz price) shall be equal to

$$_T U_x = \mathrm{E}_{\widetilde{\mathrm{P}}} \left\{ \mathrm{E}_{\mathrm{P}*} \left[H \cdot \mathbb{1}_{\{T(x) > T\}} \right] \right\} = {_T p_x} \cdot \mathrm{E}_{\mathrm{P}*}[H].$$

The problem of the insurance company is to mitigate the financial part of risk and hedge $\overline{H}$ in the financial market. However,

$$_T U_x < \mathrm{E}_{\mathrm{P}*}[H]$$

in other words, the insurance company is not able to hedge the benefit perfectly; instead, the benefit may be hedged partially.

For a fixed client age x, denote the maximum amount of capital that is going into partial hedging of $\overline{H}$ by $\widetilde{V}_0 = {_T p_x} \cdot \mathrm{E}_{\mathrm{P}*}[H]$; we can now use the results of Theorem 5.3 to derive the CVaR-optimal hedging strategy. Along with providing a way of hedging, this may be viewed as a possible way of estimating the financial exposure of contracts for given values of age. Note that by applying Theorem 5.4, we can also address the dual problem: given the financial claim and a fixed CVaR threshold, we can find the target survival probability (and hence the target age) for the contract.

Example 5.3. For the purpose of illustration, we shall use the Black–Scholes model again. Let the financial part of our model follow the Black–Scholes model with parameters $\sigma = 0.3$, $\mu = 0.09$, $r = 0.05$ and let the benefit be a call option with strike price of $K = 110$; time horizon T will vary in this example. Let the initial price of the stock be equal to $S_0 = 100$. As for the insurance part, we shall use survival probabilities listed in mortality table UP94 @2015 from McGill et al. (2004) (Uninsured Pensioner Mortality 1994 Table Projected to the Year 2015). Our objective here is to construct hedging strategies that minimize $\text{CVaR}_{0.975}$ for various values of client age and time horizon. Note that we are dealing here in a way similar to Example 5.1 for the calculation of optimal CVaR for a given amount of initial wealth. The numeric results are presented in Figure 5.3.

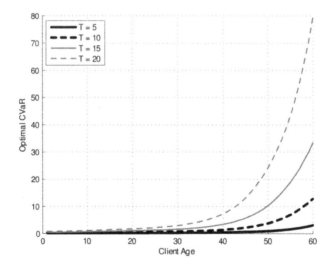

FIGURE 5.3: Optimal CVaR.

Now consider the dual problem: for a fixed CVaR threshold $\widetilde{C}$, specify the optimal client age for the equity-linked life insurance contract. If we let the underlying market be the Black–Scholes one again as above, we can do calculations similar to Example 5.2 to derive the "optimal" survival probability. Then it is just a matter of using the corresponding life table to find the optimal client age. (Note: depending on the life table, the client age may not be uniquely defined by the survival probability; in our case, we picked the highest possible value). For the following graph, we use model parameters from the previous example and survival probabilities from table UP94 @2015, McGill et al. (2004). Our findings are reflected in Figure 5.4.

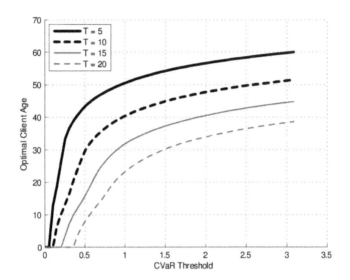

FIGURE 5.4: Optimal client age.

We finish the section showing how CVaR-hedging can be helpful in the area of financial regulation. Let us imagine that a CVaR-based capital requirement is imposed by the regulator. It is a very likely case according to the Basel 3.5 standard which suggests using CVaR instead of the well-established, but not satisfactory now, VaR-methodology. Moreover, it may lead to an important research topic arising from the practical needs of the banking industry. We propose the following approach to develop this topic.

Let γ be the amount of required regulatory capital per unit of CVaR. Assume that a contingent claim is partially hedged by using a CVaR-optimal strategy. Denote $\text{CVaR}^\alpha(v)$ as the value of the CVaR (at the confidence level α) minimizing hedging strategy which uses no more than v of the initial wealth. The option seller compares $\rho_\gamma^\alpha(0) = \gamma \cdot \text{CVaR}^\alpha(0)$ (no hedging) against $\rho_\gamma^\alpha(\tilde{V}_0) = \gamma \cdot \text{CVaR}^\alpha(\tilde{V}_0) + \tilde{V}_0$. Therefore, the ratio

$$R^\alpha(\tilde{V}_0) = \frac{\rho_\gamma^\alpha(\tilde{V}_0)}{\rho_\gamma^\alpha(0)}$$

measures the relative short-term attractiveness of partial hedging. Figure 5.5 shows the ratio $R^\alpha(\tilde{V}_0)$ for varying levels of γ and $\tilde{V}_0$ in the Black–Scholes model of Example 5.1. The regions of the graphs where the aforesaid ratio is less than 1 correspond to the case where the option seller can reduce the required regulatory capital by a value exceeding the initial cost of hedging, while at the same time reducing the total risk exposure CVaR-wise.

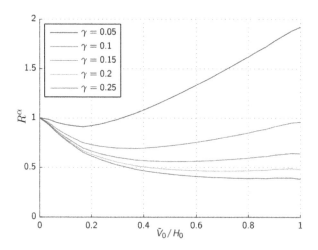

FIGURE 5.5: Performance of the CVaR-optimal hedging strategy at the 99% confidence level with varying of the initial wealth and the capital requirement ratio in the Black–Scholes model.

5.3 CVaR-hedging in the regime-switching telegraph market model

The present section focuses on the two-state telegraph market model which is a particular case of a complete regime-switching market model featuring jump dynamics of the risky asset price process. We present explicit solutions for the problems of option pricing and CVaR-optimal partial hedging in the framework of the two-state telegraph market model and illustrate the derivation of the CVaR-optimal hedging strategy numerically.

The telegraph market model can be informally described as a complete market model in which the dynamics of the risky asset features jumps and regime switching. The model can be viewed as a generalization of the pure-jump version of the Merton's model that is arbitrage-free and preserves market completeness under some restrictions on model parameters. In addition, it can be shown that the telegraph market model converges to the Black–

Scholes model in distribution under proper scaling, hence it can be used to approximate the lognormal stock price behavior.

For the sake of illustration, let us briefly compare the properties of the telegraph market model to those of the well-known Cox–Ross–Rubinstein binomial model. Both models feature jump dynamics, market completeness and convergence of the stock price process to a lognormal process in distribution. However, the key difference between the two is that in the binomial model the jumps have fixed timing and random size, while in the telegraph market model, the size of the jumps is pre-determined but they occur at random.

Consider a simplified version of the Merton's model with no diffusion term and a constant jump impact:

$$\begin{cases} dB_t = rB_t dt, \\ dS_t = S_t(m dt + \sigma dN_t), \end{cases} \quad (5.32)$$

where S_t and B_t are the prices of the stock and the bond, respectively, N_t is a Poisson process, $r \geqslant 0$ is the risk-free interest rate, $\sigma > 0$ is the constant jump impact and m is the constant drift. It is easy to verify that this model represents a complete arbitrage-free market; the risk comes only from the uncertainty in timing of jumps in the Poisson process, with the size of jumps known in advance.

How can this model be generalized without violating completeness? If we simply allow the jump size to be random, a new source of randomness will be introduced which will render the model incomplete. Instead, we can let the jump size change in a predictable fashion. Let the time between jumps be determined by a family $\{\tau_i\}_{i \in \mathbb{N}}$ of independent exponential random variables with intensities $\{\lambda_i\}_{i \in \mathbb{N}}$, and let $\{h_i\}_{i \in \mathbb{N}}$ be an arbitrary deterministic sequence specifying the size of jumps. Define the moment of the n-th jump as

$$T_n = \sum_{i=1}^{n} \tau_i,$$

let the log-price process $Y_t = \ln(S_t)$ follow

$$Y_t = mt + \sum_{i \in \mathbb{N},\, T_i \leqslant t} h_i,$$

and consider the market

$$\begin{cases} dB_t = rB_t dt, \\ dS_t = S_t dY_t. \end{cases} \quad (5.33)$$

Similar to (5.32), the market model defined by (5.33) also contains a single source of randomness: the timing of the next jump is random while the size of the jump is known in advance. However, unless certain restrictions are imposed, this market model admits arbitrage. For the sake of illustration, we shall provide a simple example.

Example 5.4. Let
$$(h_1, h_2) = (-h, h), \quad (\lambda_1, \lambda_2) = (1, 1), \quad r = 0, \quad m = \text{const} > 0, \quad h > 0.$$

In this setting, consider the following strategy on $t \in [0, 1]$:

- If a negative jump occurs at time $t = t_0 < 1$, borrow S_{t_0} in cash and buy one unit of stock. Sell the stock at time $t = t_1 \wedge 1$ and repay the debt of S_{t_0}, where $t = t_1$ is the time of the next (positive) jump;
- If no negative jumps occur on $[0, 1]$, do not do anything.

Evidently, the strategy presented above is an arbitrage strategy: if there are no negative jumps on $[0, 1]$, the profit is zero, otherwise there is a positive gain since $S_{t_1 \wedge 1} > S_{t_0}$. In the case when the drift is negative or both jumps are negative, it is straightforward to construct similar examples.

To ensure the no-arbitrage property in model (5.33) with alternating jump sizes, the drift should switch sign and/or magnitude at the time of the jump:

$$d(t) = \int_0^t \sum_{i \in \mathbb{N}} d_i \cdot \mathbb{1}_{[T_{i-1}, T_i)}(s) ds,$$

where d_i should be constrained in such way that the model does not allow arbitrage opportunities. Processes of this type are known as telegraph processes.

To summarize, the telegraph market model is a generalization of the Poisson market model obtained by introducing regime switching, where only the timing of regime switching is uncertain.

In what follows, we consider the simplest case of a telegraph process with two alternating states. This particular case is important since it inherits all of the characteristic features of a general telegraph model, yet at the same time it allows for quite simple semi-explicit solutions for the problem of option pricing and hedging.

Let $\sigma(t) \in \{1, 2\}$ be a continuous time Markov chain process with Markov generator

$$L_\sigma = \begin{pmatrix} -\lambda_1 & \lambda_1 \\ \lambda_2 & -\lambda_2 \end{pmatrix}. \tag{5.34}$$

Process $\sigma(t)$ represents the current state in the telegraph market model. Without loss of generality, we assume that $\sigma(0) = 1$.

Define the telegraph process X_t and the jump process J_t:

$$X_t = \int_0^t c_{\sigma(s)} ds, \quad J_t = \sum_0^{N_t} h_{\sigma(T_j-)}, \tag{5.35}$$

where $c = (c_1, c_2)$ determines drift states, $h = (h_1, h_2)$ determines jump size states, and N_t denotes the number of jumps of $\sigma(t)$ up to time t.

The risk-free asset is defined by

$$dB_t = r_{\sigma(t)} B_t dt \qquad (5.36)$$

or, equivalently,

$$B_t = \exp\left(\int_0^t r_{\sigma(s)} ds\right),$$

where $r = (r_1, r_2)$ determines the states of the (nonnegative) risk-free rate.

The risky asset is defined as a telegraph market price process

$$dS_t = S_{t-} d(X_t + J_t). \qquad (5.37)$$

We can express S_t as

$$S_t = \varepsilon_t(X + J) = S_0 e^{X_t} \kappa(N_t), \qquad (5.38)$$

where $\varepsilon_t(\cdot)$ denotes the stochastic exponential and

$$\kappa(N_t) = \prod_{s \leq t} (1 + \Delta J_s), \qquad (5.39)$$

which implies

$$\kappa(2k-1) = (1+h_1)^k (1+h_2)^{k-1},$$
$$\kappa(2k) = (1+h_1)^k (1+h_2)^k,$$

for all $k \in \mathbb{N}$.

One of the most important questions regarding the two-state (regime-switching) telegraph market model (5.36)–(5.37) is the absence of arbitrage and market completeness. This topic has been investigated in detail in Melnikov and Ratanov (2008), and we shall reproduce the main results here in the form of a theorem.

Theorem 5.5. *The two-state telegraph market model is arbitrage-free if and only if*

$$\frac{r_\sigma - c_\sigma}{h_\sigma} >, \quad \sigma \in \{1, 2\}. \qquad (5.40)$$

If the telegraph market model is arbitrage-free, then it is complete, and the unique equivalent martingale measure P* *is defined by*

$$\left.\frac{d\mathrm{P}^*}{d\mathrm{P}}\right|_{\mathcal{F}_t} = Z_t = \varepsilon(X^* + J^*) = e^{X_t^*} \kappa^*(N_t),$$

where

$$\kappa^*(N_t) = \prod_{s \leq t} (1 + \Delta J_s^*),$$

X_t^* is a telegraph process with intensities λ_σ defined in (5.34) and drift

$$c_\sigma^* = \lambda_\sigma + \frac{c_\sigma - r_\sigma}{h_\sigma}, \quad \sigma \in \{1,2\},$$

and J_t^* is a jump process with jumps

$$h_\sigma^* = \frac{r_\sigma - c_\sigma}{h_\sigma \lambda_\sigma} - 1, \quad \sigma \in \{1,2\}.$$

The discounted price process $\tilde{S}_T = B_t^{-1} S_t$ is a telegraph market process under P^* with drift c_σ, jumps h_σ and modified intensities

$$\lambda_\sigma^* = \frac{r_\sigma - c_\sigma}{h_\sigma} > 0.$$

We shall now turn to the problems of pricing and partial hedging of contingent claims in the setting of the telegraph market model. Consider a contingent claim of European type with maturity T and discounted payoff $H \in \mathcal{L}^1(P)$ at time $t = T$.

First, we would like to be able to compute the unique fair price of the contingent claim, i.e., the risk-neutral expected value of the discounted payoff:

$$H_0 = E^*(H),$$

where E^* denotes the expectation with respect to the unique equivalent martingale measure P^* defined by (5.40).

Second, we are interested in the solution to the following partial hedging problem. Assume that the amount of the available initial wealth $\tilde{V}_0$ is strictly bounded from above by the fair price H_0, then the perfect hedging strategy is no longer attainable, and our objective is to find a hedging strategy that minimizes the total conditional value-at-risk (CVaR) while requiring no more than $\tilde{V}_0$ of the initial capital. CVaR, also known as expected shortfall (ES), is a coherent quantile risk measure (see, Section 5.1) and can be defined as

$$\text{CVaR}^\alpha(L) = \frac{1}{1-\alpha}\left(E(\mathbb{1}_{\{L \geq L^{(\alpha)}\}} L) + L^{(\alpha)}(1 - \alpha - P(L \geq L^{(\alpha)}))\right)$$

or

$$\text{CVaR}^\alpha(L) = \frac{1}{1-\alpha} \int_\alpha^1 L^{(p)} dp = \frac{1}{1-\alpha} \int_\alpha^1 \text{VaR}^{(p)} dp,$$

where L is a random variable representing loss, $\alpha \in (0,1)$ is a predetermined confidence level, $L^{(\alpha)} = \inf\{x \in \mathbb{R} : P(L \leq t) > \alpha\}$ is the upper α-quantile of L and $\text{VaR}^\alpha(L) = L^{(\alpha)}$ is the value-at-risk of level α associated with the loss distribution L.

In the context of our partial hedging problem, a hedging strategy is represented by a tuple (V_0, ξ), where $V_0 > 0$ is the initial wealth and $\xi = (\xi_t)_{t \in [0,T]}$

is a predictable process which indicates the holding in the stock at each point of time. For each strategy (V_0, ξ) the corresponding discounted value process V_t is defined as

$$V_t = V_0 + \int_0^t \xi_u d\widetilde{S}_u,$$

where $\widetilde{S}_t = B_t^{-1} S_t$, with B_t and S_t defined by (5.36) and (5.37). We shall restrict ourselves to only considering strategies that are admissible in the sense that $V_t \geqslant 0$ for all $t \in [0, T]$, a.s. The loss functional L at the terminal moment of time $t = T$ in the CVaR definition then becomes

$$L(V_0, \xi) = H - V_T = H - V_0 - \int_0^T \xi_u d\widetilde{S}_u,$$

and the problem of interest can be stated as

$$\begin{cases} \mathrm{CVaR}^\alpha(V_0, \xi) \to \min_{(V_0, \xi) \in \mathcal{A}}, \\ V_0 \leqslant \widetilde{V}_0. \end{cases} \quad (5.41)$$

where $\mathcal{A}$ denotes the set of all admissible hedging strategies.

It turns out that in a complete market case the solution to problem (5.41) can be derived semi-explicitly by applying the Neyman–Pearson lemma to a family of subproblems, (see Section 5.1 for full details). Concretely, the optimal strategy $(\widetilde{V}_0, \widetilde{\xi})$ for the problem of CVaR minimization (5.41) is a perfect hedge for the modified contingent claim $\widetilde{H}(\widetilde{z}) = (H - z)^+ \widetilde{\varphi}(\widetilde{z})$, where $\widetilde{\varphi}(z)$ is the optimal randomized test defined by

$$\widetilde{\varphi}(z) = \mathbb{1}_{\{\frac{dP}{dP^*} > \tilde{a}(z)\}} + \gamma(z) \mathbb{1}_{\{\frac{dP}{dP^*} = \tilde{a}(z)\}}, \quad (5.42)$$

where

$$\tilde{a}(z) = \inf\left\{a \geqslant 0 : \mathrm{E}^*\left((H-z)^+ \cdot \mathbb{1}_{\{\frac{dP}{dP^*} > a\}}\right) \leqslant \widetilde{V}_0\right\},$$

$$\gamma(z) = \frac{\widetilde{V}_0 - \mathrm{E}^*\left((H-z)^+ \mathbb{1}_{\{\frac{dP}{dP^*} > \tilde{a}(z)\}}\right)}{\mathrm{E}^*\left((H-z)^+ \mathbb{1}_{\{\frac{dP}{dP^*} = \tilde{a}(z)\}}\right)}.$$

$\widetilde{z}$ is the point of global function

$$c(z) = \begin{cases} z + \frac{1}{1-\alpha} \cdot \mathrm{E}\left((H-z)^+(1 - \widetilde{\varphi}(z))\right), & \text{for } z < z^*, \\ z, & \text{for } z \geqslant z^* \end{cases} \quad (5.43)$$

on interval $z \in [0, z^*]$ is the real root of equation

$$\widetilde{V}_0 = \mathrm{E}^*\left((H - z^*)^+\right). \quad (5.44)$$

Third, we consider the problem of minimizing the initial wealth required for construction of a hedging strategy over all admissible strategies (V_0, ξ) under the condition that conditional value-at-risk of a given confidence level α does not exceed a predefined threshold $\tilde{C}$:

$$\begin{cases} V_0 \to \min_{(V_0, \xi) \in \mathcal{A}}, \\ \mathrm{CVaR}^\alpha(V_0, \xi) \leq \tilde{C}. \end{cases} \quad (5.45)$$

Similar to the problem of CVaR minimization, a semi-explicit solution can be obtained via the Neyman–Pearson approach (see Section 5.1): the optimal strategy $(\tilde{V}_0, \tilde{\xi})$ for the problem of hedging costs minimization (5.45) is

(a) A perfect hedge for the contingent claim $(H - \tilde{z})^+ (1 - \tilde{\varphi}(\tilde{z}))$ if the following two conditions are satisfied:

$$\mathrm{E}(H) > \tilde{C}_0, \quad \mathrm{E}\left((H - \tilde{C})^+\right) > 0, \quad (5.46)$$

where $\tilde{C}_z = (\tilde{C} - z)(1 - \alpha)$, $\tilde{\varphi}(z)$ is the optimal randomized test defined by

$$\tilde{\varphi}(z) = \mathbb{1}_{\{\frac{dP}{dP^*} > \tilde{a}(z)\}} + \gamma(z) \mathbb{1}_{\{\frac{dP}{dP^*} = \tilde{a}(z)\}}, \quad (5.47)$$

$$\tilde{a}(z) = \inf\left\{a \geq 0 : \mathrm{E}\left((H - z)^+ \cdot \mathbb{1}_{\{\frac{dP}{dP^*} > a\}}\right) \leq \tilde{C}_z\right\},$$

$$\gamma(z) = \frac{\tilde{C}_z - \mathrm{E}\left((H - z)^+ \mathbb{1}_{\{\frac{dP}{dP^*} > \tilde{a}(z)\}}\right)}{\mathrm{E}\left((H - z)^+ \mathbb{1}_{\{\frac{dP}{dP^*} = \tilde{a}(z)\}}\right)}.$$

and $\tilde{z}$ is a point of minimum of function

$$d(z) = \mathrm{E}^*\left((H - z)^+ (1 - \tilde{\varphi}(z))\right) \quad (5.48)$$

on the interval $-\infty < z \leq \tilde{C}$;

(b) A passive trading strategy:

$$\tilde{V}_t = 0, \quad t \in [0, T],$$

if any of the two conditions in (5.46) are not satisfied.

It is straightforward to see that computing H_0, $\tilde{\varphi}(z)$ in (5.42) and (5.47), $c(z)$ in (5.43), z^* in (5.44) and $d(z)$ in (5.48) involves evaluating expectations of the general form

$$\mathrm{E}_t(f, a) = \mathrm{E}(f(S_t, B_t) \cdot \mathbb{1}_{\{Z_t < a\}}) \quad (5.49)$$

and

$$\mathrm{E}_t^*(f, a) = \mathrm{E}^*(f(S_t, B_t) \cdot \mathbb{1}_{\{Z_t < a\}}) \quad (5.50)$$

for arbitrary functions $f : [0, \infty) \times \mathbb{N}_0 \mapsto \mathbb{R}$ and $a \in (-\infty, +\infty]$, where $\mathbb{N}_0$ is the set of natural numbers including zero. We shall investigate these expectations in great detail.

First, note that
$$E_t^*(f, a) = E_t(f \cdot Z_t, a),$$
hence it is sufficient to only consider the expectation $E_t(f, a)$ under the real-world measure P.

We shall now try and express processes B_t, S_t and Z_t in terms of X_t and N_t. By the definition of B_t,

$$B_t = \exp\left(\int_0^t r_{\sigma(s)} ds\right) = \exp\left(\frac{r_2 - r_1}{c_2 - c_1} X_t + \frac{c_2 r_1 - c_1 r_2}{c_2 - c_1} t\right). \quad (5.51)$$

Process S_t can be related to X_t and N_t via (5.38), and according to Theorem 5.5,
$$Z_t = e^{X_t^*} \kappa^*(N_t). \quad (5.52)$$

In order to express Z_t in terms of X_t and N_t, we shall use the fact which states that two telegraph processes based on the same process $\sigma(t)$ are linearly related. More specifically,
$$X_t^* = \frac{c_2^* - c_1^*}{c_2 - c_1} X_t + \frac{c_2 c_1^* - c_1 c_2^*}{c_2 - c_1} t,$$
which in conjunction with (5.52) yields

$$Z_t = \exp\left(\frac{c_2^* - c_1^*}{c_2 - c_1} X_t + \frac{c_2 c_1^* - c_1 c_2^*}{c_2 - c_1} t\right) \kappa^*(N_t). \quad (5.53)$$

We can now substitute (5.38), (5.51) and (5.53) into (5.49):
$$E_t(f, a) = E(f(S_t, B_t) \cdot \mathbf{1}_{\{Z_t < a\}}) = E(g(X_t, N_t)).$$

On condition of $\{N_t = n\}$, we obtain

$$E(g(X_t, N_t)) = \sum_{\mathbb{N}_0} E\left(g(X_t, n) \cdot \mathbf{1}_{\{N_t = n\}}\right) = \sum_{\mathbb{N}_0} \int_\mathbb{R} g(x, n) p_n(t, x) dx, \quad (5.54)$$

where $p_n(t, x)$ is the corresponding conditional density

$$p_n(t, x) = \frac{d}{dx} P\left(\{X_t \leq x\} \cap \{N_t = n\}\right).$$

Therefore, we can reduce evaluating expectations of the form (5.49) and (5.50) to a summation of one-dimensional integrals with respect to densities $p_n(t, x)$. Theorem 5.6 provides the recursive relationship for these conditional densities so that the expectation can be computed explicitly.

Theorem 5.6. *In the two-state telegraph market model with processes X_t and N_t defined by (5.35), for an arbitrary function $g : [0, \infty) \times \mathbb{N}_0 \mapsto \mathbb{R}$,*

$$E(g(X_t, N_t)) = \sum_{\mathbb{N}_0} \int_{\mathbb{R}} g(x, n) p_n(t, x) dx,$$

where for all $t \geq 0$ and $x \in \mathbb{R}$,

$$p_0(t, x) = e^{-\lambda_1 t} \delta(x - c_1 t),$$

and for all $k \in \mathbb{N}$,

$$p_{2k-1}(t, x) = \frac{\lambda_1 \left(\phi_1(t, x)\phi_2(t, x)\right)^{k-1}}{|c_2 - c_1|((k-1)!)^2} e^{-\phi_1(t,x) - \phi_2(t,x)}, \tag{5.55}$$

$$p_{2k}(t, x) = \frac{p_{2k-1}(t, x)\phi_2(t, x)}{k}, \tag{5.56}$$

and

$$\phi_1(t, x) = \lambda_1 \frac{c_2 t - x}{c_2 - c_1}, \quad \phi_1(t, x) = \lambda_2 \frac{x - c_1 t}{c_2 - c_1},$$

and $x \in (c_1 t \wedge c_2 t, c_1 t \vee c_2 t)$.

Proof. Denote by T_j the time of the j-th jump of $\sigma(t)$ and denote by $\tau_j = T_j - T_{j-1}$ the length of time between two successive jumps. Since $\sigma(t)$ is a continuous-time Markov chain, random variables τ_j are exponentially distributed:

$$\tau_{2k+1} \sim \varepsilon(\lambda_1), \quad \tau_{2k} \sim \varepsilon(\lambda_2),$$

where $\varepsilon(\lambda)$ is the exponential distribution with parameter λ.

Denote

$$S_k^{\text{odd}} = \tau_1 + \tau_3 + \ldots + \tau_{2k-1},$$
$$S_k^{\text{even}} = \tau_2 + \tau_4 + \ldots + \tau_{2k}.$$

Then

$$S_k^{\text{odd}} \sim \Gamma(k, \lambda_1), \quad S_k^{\text{even}} \sim \Gamma(k, \lambda_2),$$

where $\Gamma(\alpha, \beta)$ is the gamma distribution with parameters α and β.

Without loss of generality, we assume that $c_2 > c_1$; the other case can be handled similarly. Note that the drift of X_t is equal to c_1 when $T_{2j} < t < T_{2j+1}$, and it is equal to c_2 when $T_{2j-1} < t < T_{2j}$.

Hence, if $N_t = 2k$, we have $T_{2k} < t < T_{2k+1}$ and $T_{2k} = S_k^{\text{odd}} + S_k^{\text{even}}$, thus

$$X_t = c_1 S_k^{\text{odd}} + c_2 S_k^{\text{even}} + c_1(t - T_{2k}) = (c_2 - c_1) S_k^{\text{even}} + c_1 t.$$

Similarly, if $N_t = 2k + 1$,

$$X_t = c_1 S_{k+1}^{\text{odd}} + c_2 S_k^{\text{even}} + c_2(t - T_{2k+1}) = (c_1 - c_2) S_{k+1}^{\text{odd}} + c_2 t.$$

By using the identity
$$\{N_t = 2k\} = \{T_{2k} \leqslant t < T_{2k} + \tau_{2k+1}\},$$
we conclude that
$$p_{2k}(t,x) = \mathrm{P}\left(\{(c_2 - c_1)S_k^{\mathrm{even}} < x - c_1 t\} \right.$$
$$\left. \cap \{S_k^{\mathrm{odd}} + S_k^{\mathrm{even}} < t < S_k^{\mathrm{odd}} + S_k^{\mathrm{even}} + \tau_{2k+1}\}\right).$$

Finally, since S_k^{odd}, S_k^{even} and τ_{2k+1} are independent random variables with distributions $\Gamma(k, \lambda_1)$, $\Gamma(k, \lambda_2)$ and $\varepsilon(\lambda_1)$, respectively, we obtain

$$p_{2k}(t,x) = \frac{\lambda_1^{k+1} \lambda_2^k}{((k-1)!)^2} \int_0^{\frac{x-c_1 t}{c_2 - c_1}} dz_2 \int_0^{t-z_2} dz_1 \int_{t-z_1-z_2}^{\infty} dz_3 (z_1 z_2)^{k-1} e^{-\lambda_1(z_1+z_3) - \lambda_2 z_2}.$$
(5.57)

By differentiating both sides of (5.57) with respect to x and computing the double integral explicitly, we arrive at (5.56). The expression (5.55) for the density at the odd indices can be derived in a similar way. □

Example 5.5. To illustrate how conditional value-at-risk can be minimized numerically in the setting of the two-state telegraph market model, we shall derive the CVaR-optimal hedging strategy (along with the hedging costs minimizing strategy) numerically for a European call option with the strike price of $K = 100$ and time to maturity $T = 0.25$ in the telegraph market model with parameters $c = (-0.5, 0.5)$, $\lambda = (5, 5)$, $r = (0.07, 0.07)$, $h = (0.5, -0.35)$, $S_0 = 100$.

The process of finding the CVaR-optimal strategy involves computing the expected value of various functions a large number of times. One possible way of evaluating the integral in (5.54) efficiently is to consider a fixed grid

$$x_i = x_{\min} + \frac{i}{2N^{(x)}} \cdot (x_{\max} - x_{\min}), \quad i = 0, 1, \ldots, 2N^{(x)},$$

$$n_j = j, \quad j = 0, 1, \ldots, N^{(n)},$$

where $x_{\min} = c_1 T \wedge c_2 T$, $x_{\max} = c_1 T \vee c_2 T$, and approximate the expectation by partitioning the interval $(x_{\min}, x_{\max})$ into $N^{(x)}$ parts:

$$\sum_{N_0} \int_{\mathbb{R}} g(x,n) p_n(t,x) dx \approx \sum_{i=0}^{2N^{(x)}} \sum_{j=0}^{N^{(n)}} g(x_i, n_j) p_{n_j}(T, x_i) \zeta_i,$$

where values $p_{n_j}(T, x_i)$ are computed in advance, and ζ_i are the Simpson's method weights: $\zeta_0 = \zeta_{N^{(x)}} = \frac{1}{3}$, $\zeta_{2k} = \frac{2}{3}$, $\zeta_{2k+1} = \frac{4}{3}$.

Figure 5.6 shows the minimal CVaR that can be attained by using the CVaR-optimal hedging strategy in the telegraph market model for various values of the initial wealth and confidence level. In Figure 5.7, we present the numerical solution for the problem of hedging costs minimization in the telegraph market model.

CVaR-Hedging: Theory and applications 135

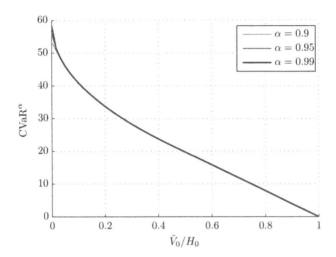

FIGURE 5.6: CVaR of the optimal hedging strategy at confidence levels of 90%, 95% and 99% for varying levels of initial wealth in the telegraph market model.

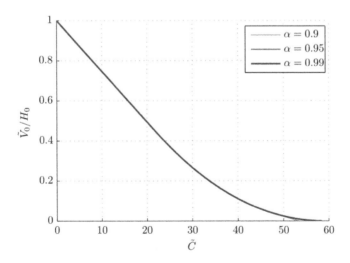

FIGURE 5.7: Initial wealth of the optimal hedging strategy for varying levels of the CVaR threshold at confidence levels of 90%, 95% and 99% in the telegraph market model.

Chapter 6

Defaultable securities and equity-linked life insurance contracts

6.1 Multiple defaults and defaultable claims in the Black–Scholes model 137
6.2 Efficient hedging in defaultable market: Essence of the technique and main results 143
6.3 Application to equity-linked life insurance contracts: Brennan–Schwartz approach vs. the superhedging 158

6.1 Multiple defaults and defaultable claims in the Black–Scholes model

We know that in a complete market, starting with a large enough initial capital, one can construct a perfect hedge for every contingent claim. However, if the market is incomplete. the initial cost of superhedging is too high. As we know the defaultable markets usually turn into incomplete markets, and the default time which is represented by a random time cannot be hedged by investing in the available assets in the market. This issue makes superhedging too expensive in defaultable markets. Therefore, we are forced to introduce new measures of risk and start investing with a smaller initial capital than the superhedging cost. But the high cost of superhedging is not the only reason that makes efficient hedging interesting. It is true that the perfect hedge or superhedge eliminates risk but it eliminates opportunities too. There are financial institutions that seek out risk; financial institutions as insurance companies expose themselves intentionally to risk and exploit risk to generate value.

We consider a financial model consisting of two assets B and S, defined by their price processes $(B_t)_{0 \leq t \leq T}$ and $(S_t)_{0 \leq t \leq T}$. Let us call this model the (B, S)-market and assume its price evolution as follows

$$dS_t = S_t(m_t dt + \sigma_t dW_t), \quad S_0 \in (0, \infty)$$
$$dB_t = B_t r_t dt, \quad B_0 = 1 \tag{6.1}$$

for $t \in [0, T]$. $(r_t)_{0 \leq t \leq T}$ is the risk-free interest rate of our bank account B, volatility and appreciation rate of S as the risky asset are given by $\sigma > 0$ and

m, respectively, and $(W_t)_{0 \leq t \leq T}$ is a standard Brownian motion on the complete probability space $(\Omega, \mathbb{F} = (\mathcal{F}_t)_{0 \leq t \leq T} \subseteq \mathcal{G}, \mathrm{P})$. For the sake of simplicity, we assume $r \equiv 0$, and postulate that (6.1) is a complete market. In other words, there exists a unique equivalent martingale measure P* defined by

$$\frac{d\mathrm{P}^*}{d\mathrm{P}} = \rho_T^*$$

such that

$$\rho_t^* = \exp\left(-\int_0^t \frac{m_s}{\sigma_s} dW_s - \int_0^t \frac{1}{2}\left(\frac{m_s}{\sigma_s}\right)^2 ds\right) \tag{6.2}$$

for $t \in [0, T]$. In addition, $(\rho_t^*)_{0 \leq t \leq T}$ satisfies the following integrability conditions

(1) $\int_0^T \left(\frac{m_s}{\sigma_s}\right)^2 ds < +\infty$, (P-a.s.),

(2) $\mathrm{E}[\rho_T^*] = 1$.

The default times are represented by τ_i for $i = 1, \ldots, n$. They are some positive $\mathcal{G}$-measurable random variables ($\mathcal{G}$-random times) with $\mathrm{P}(\tau_i = 0) = 0$ such that $\mathrm{P}(\tau_i > t) > 0$ for all $i = 1, \ldots, n$ and $t \in [0, T]$. For $J \subseteq \{1, \ldots, n\}$, by progressively enlarging of the filtrations, define

$$\mathcal{G}_t^J = \mathcal{F}_t \vee \mathcal{H}_t^J \quad \text{and} \quad \mathbb{G}^J = (\mathcal{G}_t^J)_{0 \leq t \leq T}$$

where $\mathcal{H}_t^J := \sigma(\tau_i \wedge t; i \in J)$, for $t \in [0, T]$. If $J = \{1, \ldots, n\}$, we simply write $\mathcal{G}_t$, $\mathcal{H}_t$, and $\mathbb{G}$. The market (6.1) with default times (τ_i), $i = 1, \ldots, n$ is called defaultable and is denoted as the (B, S, τ)-market.

We make the following assumptions on the default times:

Assumption 6.1. *The default times $\{\tau_i : i = 1, \ldots, n\}$ and $(W_t)_{0 \leq t \leq T}$ are mutually independent.*

Assumption 6.2. $\mathrm{P}(\tau_i = \tau_j) = 0$, *for all $i, j = 1, \ldots, n$ and $i \neq j$.*

Let's note that under Assumption 6.1 any $\mathbb{F}$-martingale remains a $\mathbb{G}$-martingale. This guarantees no arbitrage condition in the defaultable (B, S, τ)-market.

There are some crucial processes associated with each random time τ_i for $i = 1, \ldots, n$:

- The $\mathbb{F}$-supermartingale

$$G_t^i := \mathrm{P}(\tau_i > t | \mathcal{F}_t) \tag{6.3}$$

is called the Azéma supermartingale or the survival process of τ_i with respect to $\mathbb{F}$. We assume that $G_t^i > 0$ for all $t \in [0, T]$. Due to Assumption 6.1, in our model (6.3) is simplified to $G_t^i = \mathrm{P}(\tau_i > t)$.

- For $i = 1, ..., n$, if there exists a nonnegative $\mathbb{F}$-predictable process $(\mu_t^i)_{0 \leq t \leq T}$ such that

$$G_t^i = \exp\left(-\int_0^t \mu_s^i ds\right), \quad t \in [0,T]$$

then $(\mu_t^i)_{0 \leq t \leq T}$ is called $\mathbb{F}$-intensity of the random time τ_i. This assumption on the default time is well known as the intensity hypothesis. Since $G_t^i = \mathrm{P}(\tau_i > t)$ by Assumption 6.1, for all $i = 1, \ldots, n$ the intensity μ_t^i is only a nonnegative function of the variable $t \in [0,T]$.

- If $(\mu_t^i)_{0 \leq t \leq T}$ exists then

$$M_t^i := \mathbb{1}_{\{\tau_i \leq t\}} - \int_0^{\tau_i \wedge t} \mu_s^i ds, \quad t \in [0,T]$$

is a $\mathbb{G}$-martingale.

We can interpret the intensity process as a local default rate, i.e.,

$$\mathrm{P}(\tau_i \in (t, t+dt]|\mathcal{F}_t) = \mu_t^i dt.$$

Let us define

$$\mathcal{D} = \{(\kappa_t)_{0 \leq t \leq T}: \text{ bounded, } \mathbb{G}\text{-predictable and } \kappa_t > -1 \, dt \times d\mathrm{P} \text{ a.s.}\}.$$

Keeping in mind the above assumptions and notations, one can show that the class of equivalent martingale measures for our defaultable model is given by:

$$\mathcal{Q} = \{Q^\kappa \mid \kappa := (\kappa_t^1, ..., \kappa_t^n)_{0 \leq t \leq T} \in \mathcal{D}^n\}$$

where the set $\mathcal{D}^n$ consists of all n-tuples of the elements of $\mathcal{D}$, and

$$\frac{dQ^\kappa}{d\mathrm{P}} = \rho_T^* \rho_T^\kappa, \quad \kappa \in \mathcal{D}^n$$

with

$$\rho_t^\kappa = 1 + \sum_{i=1}^n \int_0^t \kappa_s^i \rho_{s-}^{\kappa^i} dM_s^i, \quad t \in [0,T].$$

Due to Assumption 6.2, the jumps of τ_i and τ_j for $i \neq j$ do not coincide. Thus we can represent ρ^κ as follows

$$\rho_t^\kappa = \left(\prod_{i=1}^n (1 + \kappa_{\tau_i}^i \mathbb{1}_{\{\tau_i \leq t\}})\right) \exp\left(-\sum_{i=1}^n \int_0^{\tau_i \wedge t} \kappa_s^i \mu_s^i ds\right), \quad t \in [0,T]. \quad (6.4)$$

Based on the available information ($\mathbb{G}$) to the traders in the defaultable market, we give the following definition for admissible strategies.

- A $\mathbb{G}$-trading strategy is a $\mathbb{G}$-predictable process $\pi := \left(\pi_t^0, \pi_t^1\right)_{t \in [0,T]}$ such that
$$\int_0^T |\pi_t^0| dt < \infty, \text{ and } \int_0^T \left(\pi_t^1 S_t\right)^2 dt < \infty \text{ P-a.s.}$$

- At time $t \in [0,T]$, the value process associated to $\left(\pi_t^0, \pi_t^1\right)_{t \in [0,T]}$ is defined by
$$V_t := \pi_t^0 + \pi_t^1 S_t.$$

- For a given initial value $v_0 \geqslant 0$, the trading strategy π is called self-financing if its corresponding value process satisfies
$$V_t = v_0 + \int_0^t \pi_s^1 dS_s, \quad \text{(P-a.s.)}$$
for all $t \in [0,T]$.

- A self-financing strategy $(v_0, \pi_t)_{t \in [0,T]}$ is called $\mathbb{G}$-admissible, if for its corresponding value process $(V_t^{v_0,\pi})_{t \in [0,T]}$ we have
$$V_t^{v_0,\pi} \geqslant 0, \quad \text{(P-a.s.)} \ \forall t \in [0,T].$$

The set of all $\mathbb{G}$-admissible strategies with initial value v_0 is denoted by $\mathcal{A}^{\mathbb{G}}(v_0)$. In a similar way to the above, we can define $\mathcal{A}^{\mathbb{F}}(v_0)$, the $\mathbb{F}$-admissible thhe strategies with initial value v_0.

We investigate the efficient hedging problem of the defaultable contingent claims defined as
$$H_\delta := H \prod_{i=1}^n \left(\mathbb{1}_{\{\tau_i > T\}} + \delta_i \mathbb{1}_{\{\tau_i \leqslant T\}} \right) \tag{6.5}$$
where H is a nonnegative $\mathcal{F}_T$-measurable random variable and $\delta_i \in [0,1]$ is called the i-recovery rate, for $i = 1, \ldots, n$.

To formulate the problem of minimizing the shortfall risk weighted by a loss function l we give its definition more detail in comparison with Chapter 3.

A loss function l is an increasing convex function on $[0, +\infty)$ with

- $l(0) = 0$,
- $E[l(H)] < +\infty$

Additionally, we make some differentiability assumptions about l.

Assumption 6.3. *The function l has the following properties:*

- $l \in C^1(0, +\infty)$,

- l' is strictly increasing on $(0, +\infty)$,
- $l'(0+) = 0$ and $l'(+\infty) = +\infty$.

Due to Assumption 6.3, the inverse function of l' exists and is denoted by I, $I := (l')^{-1}$.

In this framework, we apply the superhedging techniques to hedge H_δ in the incomplete market (B, S, τ). The initial cost of superhedging is defined by

$$U_0 := \inf \{u \geq 0 : V_T^{u,\pi} \geq H_\delta, \text{ for some } \pi \in \mathcal{A}^{\mathbb{G}}(u)\}.$$

Equivalently, we can show that

$$U_0 = \sup_{Q \in \mathcal{Q}} E^Q[H_\delta]$$

where $\mathcal{Q}$ is the class of all martingale measures for S with respect to $(\Omega, \mathbb{G}, P)$. We define

$$\bar{X}_t := \operatorname*{ess\,sup}_{Q \in \mathcal{Q}} E^Q[H_\delta | \mathcal{G}_t]$$

for $t \in [0,T]$. Then it is known that for some $\bar{\pi} \in \mathcal{A}^{\mathbb{G}}(U_0)$ and an increasing optional process $\bar{C}$ with $\bar{C}_0 = 0$, we have

$$\bar{X}_t = U_0 + \int_0^t \bar{\pi}_s dS_s - \bar{C}_t. \tag{6.6}$$

This decomposition is called the optional decomposition of H_δ defined by (6.5).

Using the structure of $\mathcal{Q}$ we provide a useful representation for U_0. In the following, the expectation $E[\rho_T^* H]$ is denoted by $E^*[H]$.

Lemma 6.1. *If $E^*[H] < +\infty$ then $U_0 = E^*[H]$.*

Proof. For simplicity of notation, we suppose that $n = 2$, $\delta_1 \neq 0$ and $\delta_2 = 0$, and other cases can be treated similarly. In this case, $\{H_\delta > 0\} = \{H > 0\} \cap \{\tau_2 > T\}$ and

$$U_0 = \sup_{\kappa \in \mathcal{D}^2} E[\rho_T^* \rho_T^\kappa H_\delta]$$

$$= \sup_{\kappa \in \mathcal{D}^2} \left\{ E\left[\rho_T^* \rho_T^\kappa H \mathbb{1}_{\{\tau_2 > T\} \cap \{\tau_1 > T\}}\right] \right.$$

$$\left. + \delta_1 E\left[\rho_T^* \rho_T^\kappa H \mathbb{1}_{\{\tau_2 > T\}} (1 - \mathbb{1}_{\{\tau_1 > T\}})\right] \right\} \tag{6.7}$$

$$= \sup_{\kappa \in \mathcal{D}^2} \left\{ \delta_1 E\left[\rho_T^* \rho_T^\kappa H \mathbb{1}_{\{\tau_2 > T\}}\right] \right.$$

$$\left. + (1 - \delta_1) E\left[\rho_T^* \rho_T^\kappa H \mathbb{1}_{\{\tau_2 > T\} \cap \{\tau_1 > T\}}\right] \right\}.$$

Now, let us consider $\kappa = (\kappa^1, \kappa^2) \in \mathcal{D}^2$ with $\kappa^1 \in \mathcal{D}$ arbitrary but $\kappa^2 \in \mathcal{D}$ constant such that $\kappa^2 \searrow -1$. Then using (6.4) we obtain from (6.7) that

$$U_0 \geq \lim_{\kappa^2 \searrow -1} \left\{ \delta_1 \mathrm{E}\left[\rho_T^* \rho_T^{\kappa^1} \rho_T^{\kappa^2} H \mathbb{1}_{\{\tau_2 > T\}}\right] \right.$$
$$\left. + (1-\delta_1)\mathrm{E}\left[\rho_T^* \rho_T^{\kappa^1} \rho_T^{\kappa^2} H \mathbb{1}_{\{\tau_2 > T\} \cap \{\tau_1 > T\}}\right] \right\}$$
$$= \delta_1 \mathrm{E}\left[\rho_T^* \rho_T^{\kappa^1} \exp\left(\int_0^T \mu_s^2 ds\right) H \mathbb{1}_{\{\tau_2 > T\}}\right] \qquad (6.8)$$
$$+ (1-\delta_1) \mathrm{E}\left[\rho_T^* \rho_T^{\kappa^1} \exp\left(\int_0^T \mu_s^2 ds\right) H \mathbb{1}_{\{\tau_2 > T\} \cap \{\tau_1 > T\}}\right].$$

We work on each term of the right-hand side of (6.8), separately:

$$\mathrm{E}\left[\rho_T^* \rho_T^{\kappa^1} H \mathbb{1}_{\{\tau_2 > T\}}\right] = \mathrm{E}\left[\rho_T^{\kappa^1} \mathrm{E}\left[\rho_T^* H \mathbb{1}_{\{\tau_2 > T\}} | \mathcal{F}_T \vee \mathcal{H}_T^2\right]\right]. \qquad (6.9)$$

Let's note that $\left(\rho_t^{\kappa^1}\right)_{t \in [0,T]}$ and $\left(\mathrm{E}\left[\rho_T^* H \mathbb{1}_{\{\tau_2 > T\}} | \mathcal{F}_t \vee \mathcal{H}_t^2\right]\right)_{t \in [0,T]}$ are two orthogonal $(\mathbb{G}, \mathrm{P})$-local martingales. This implies that their product is a $(\mathbb{G}, \mathrm{P})$-local martingale as well. By considering $(H \wedge m)_{m \geq 1}$ and then using the monotone convergence theorem as $m \to +\infty$, (6.9) becomes

$$\rho_0^{\kappa^1} \mathrm{E}\left[\rho_T^* H \mathbb{1}_{\{\tau_2 > T\}} | \mathcal{F}_0 \vee \mathcal{H}_0^2\right] = \exp\left(-\int_0^T \mu_s^2 ds\right) \mathrm{E}^*[H]. \qquad (6.10)$$

On the other hand, if we first apply (6.10) in (6.8), and then choose $\kappa^1 \in \mathcal{D}$ constant such that $\kappa^1 \searrow -1$ in the second term. Then

$$U_0 \geq \delta_1 \mathrm{E}^*[H] + (1-\delta_1)\mathrm{E}^*[H] = \mathrm{E}^*[H], \qquad (6.11)$$

we used the fact that $\mathcal{F}_T$ and $\{\tau_i\}_{i=1,2}$ are independent and $\mathrm{P}\left(\bigcap_{i=1}^2 \{\tau_i > T\}\right) = \exp\left(-\sum_{i=1}^2 \int_0^T \mu_s^i ds\right)$.

To prove the reverse inequality in (6.11), notice that since $\kappa > -1$, we have

$$\rho_T^\kappa \mathbb{1}_{\{\tau_2 > T\}} = \rho_T^{\kappa^1} \exp\left(-\int_0^T \kappa_s^2 \mu_s^2 ds\right) \mathbb{1}_{\{\tau_2 > T\}}$$
$$\leq \rho_T^{\kappa^1} \exp\left(\int_0^T \mu_s^2 ds\right) \mathbb{1}_{\{\tau_2 > T\}}, \qquad (6.12)$$

and

$$\rho_T^{\kappa} \mathbb{1}_{\{\tau_2 > T\} \cap \{\tau_1 > T\}} \leq \exp\left(\sum_{i=1}^{2} \int_0^T \mu_s^i \, ds\right) \mathbb{1}_{\{\tau_2 > T\} \cap \{\tau_1 > T\}} \quad (6.13)$$

for any $\kappa = (\kappa^1, \kappa^2) \in \mathcal{D}^2$. By (6.7) and inequalities (6.12) and (6.13), we get

$$U_0 \leq \delta_1 \mathrm{E}[\rho_T^* \rho_T^{\kappa^1} H] + (1 - \delta_1) \mathrm{E}^*[H]$$
$$= \delta_1 \mathrm{E}^*[H] + (1 - \delta_1) \mathrm{E}^*[H] = \mathrm{E}^*[H]. \quad (6.14)$$

To get the first equality in (6.14), we need to repeat the arguments applied to $\mathrm{E}\left[\rho_T^* \rho_T^{\kappa^1} H \mathbb{1}_{\{\tau_2 > T\}}\right]$ in (6.9) and (6.10) for the case of $\mathrm{E}[\rho_T^* \rho_T^{\kappa^1} H]$. □

6.2 Efficient hedging in defaultable market: Essence of the technique and main results

Clearly, the client is not willing to pay $\mathrm{E}^*[H]$ for buying H_δ. One can buy H in the default-free market for this price and receive H without the risk of default. To offer a competitive price which is reasonable for the client, the premium charged by the company should be less than $\mathrm{E}^*[H]$. However, if the premium is less than $E^*[H]$ there is a possibility of shortfall for the company (still there is the possibility of payment H). This naturally leads us to the problem of minimizing the shortfall risk with an initial capital $\tilde{u} < \mathrm{E}^*[H]$. In this case, we consider the problem of minimizing the expectation of the shortfall risk weighted by a loss function. More precisely, we want to solve the following efficient hedging problem

$$\min_{\substack{\pi \in \mathcal{A}^{\mathbb{G}}(v) \\ v \leq \tilde{u}}} \mathrm{E}\left[l\left((H_\delta - V_T^{v,\pi})^+\right)\right]. \quad (6.15)$$

As we know from Section 3.1, the solution to problem (6.15) can be obtained with the help of the Föllmer-Leukert approach (see (3.2)–(3.3) and Lemma 3.1). In our situation here, we can describe this method as follows. Let $\mathcal{R} = \{\varphi : \Omega \to [0,1], \varphi \in \mathcal{G}_T\}$. Then there exists $\tilde{\varphi} \in \mathcal{R}$ that solves the problem

$$\inf_{\varphi \in \mathcal{R}} \mathrm{E}[l(1-\varphi)H_\delta] \quad (6.16)$$

under the constraint

$$\sup_{\kappa \in \mathcal{D}^n} \mathrm{E}[\rho_T^* \rho_T^{\kappa} \varphi H_\delta] \leq \tilde{u}.$$

Since $l' > 0$, then any two solutions of (6.16) are equal (a.s) on the set $\{H_\delta > 0\}$, and one can assume that $\tilde\varphi = 1$ on the set $\{H_\delta = 0\}$. The hedging strategy $(\tilde u, \tilde\pi)$ of the modified claim $\tilde\varphi \cdot H_\delta$ obtained from the optional decomposition (6.6) solves (6.15).

With the help of the next lemma we construct a bridge between the above minimization problems (6.15)–(6.16) and a maximization problem. Using this lemma together with the Cvitanic–Karatzas algorithm for solving of the dual extremal problem, we can find a closed-form expression for the solution $\tilde\varphi$.

Lemma 6.2. *Let us define $\mathcal{R} := \{\varphi : \Omega \longrightarrow [0,1] \mid \varphi \in \mathcal{G}_T\}$, and $\tilde\varphi \in \mathcal{R}$ to be determined as above. Then the random variable $\tilde\varphi$ is a solution to the following maximization problem*

$$\max_{\varphi \in \mathcal{R}} E^{\bar P}[\varphi] \qquad (6.17)$$

subject to the constraint

$$\sup_{\kappa \in \mathcal{D}^n} E[\rho_T^* \rho_T^\kappa \varphi H_\delta] \leq \tilde u, \qquad (6.18)$$

where

$$\frac{d\bar P}{dP} := \begin{cases} \dfrac{l'((1-\tilde\varphi)H_\delta)H_\delta}{E[l'((1-\tilde\varphi)H_\delta)H_\delta]}, & \text{on } \{H_\delta > 0\} \cap \{\tilde\varphi \neq 1\}; \\ 0, & \text{otherwise.} \end{cases}$$

Proof. Let us define function $F : L^1(P) \longrightarrow \mathbb{R}$ as $F(\psi) := E[l((1-\psi)H_\delta)]$ for $\psi \in L^1(P)$. Then $\tilde\varphi$ minimizes F over the convex set

$$\mathcal{R}(\tilde u) := \{\varphi \in \mathcal{R} \mid \sup_{\kappa \in \mathcal{D}^n} E[\rho_T^* \rho_T^\kappa \varphi H_\delta] \leq \tilde u\} \subset L^1(P).$$

As a consequence, the following inequality holds for the Gateaux derivative of F at $\tilde\varphi$ with the increment $\varphi - \tilde\varphi$

$$\left.\frac{dF(\tilde\varphi + t(\varphi - \tilde\varphi))}{dt}\right|_{t=0} = DF(\tilde\varphi; \varphi - \tilde\varphi) \geq 0, \text{ for all } \varphi \in \mathcal{R}(\tilde u).$$

Using monotone convergence theorem, we have

$$-E\left[l'((1-\tilde\varphi)H_\delta)(\varphi - \tilde\varphi)H_\delta\right] \geq 0, \qquad (6.19)$$

for all $\varphi \in \mathcal{R}(\tilde u)$. Equation (6.19) implies the following crucial inequality

$$E\left[l'((1-\tilde\varphi)H_\delta)\tilde\varphi H_\delta\right] \geq E\left[l'((1-\tilde\varphi)H_\delta)\varphi H_\delta\right].$$

This inequality proves the optimality for $\tilde\varphi$ to the desired problem. □

Let $\mathcal{L}$ be the closed convex hull of $\{\rho_T^\kappa\}_{\kappa \in \mathcal{D}^n}$ under P-a.s. convergence. It is clear that $\mathcal{L}$ is a convex, bounded set in $L^1(P)$ such that

$$\{\rho_T^\kappa\}_{\kappa \in \mathcal{D}^n} \subseteq \mathcal{L}.$$

Now notice that

$$E^{\bar{P}}[\varphi] = E\left[\varphi\left(\frac{d\bar{P}}{dP} - z\rho_T^* L H_\delta\right)\right] + E\left[z\rho_T^* L\varphi H_\delta\right]$$
$$\leq E\left[\left(\frac{d\bar{P}}{dP} - z\rho_T^* L H_\delta\right)^+\right] + z\tilde{u}, \qquad (6.20)$$

and

$$E\left[\rho_T^* L\varphi H_\delta\right] \leq \tilde{u} \qquad (6.21)$$

for all $\varphi \in \mathcal{R}(\tilde{u})$, $L \in \mathcal{L}$ and $z > 0$. To get inequality (6.21), we applied Fatou's lemma and (6.18).

By (6.20) and (6.21), we introduce the dual problem of the primal problem (6.17)–(6.18) as follows:

$$V_*(\tilde{u}) := \inf_{\substack{z>0 \\ L\in\mathcal{L}}} \left\{ \tilde{u}z + E\left[\left(\frac{d\bar{P}}{dP} - z\rho_T^* L H_\delta\right)^+\right]\right\}. \qquad (6.22)$$

Cvitanic and Karatzas adapted the techniques of nonsmooth convex analysis to prove that there exists a solution $(\tilde{z}, \tilde{L}) \in \mathbb{R}^+ \times \mathcal{L}$ to this dual problem. Using inequality (6.20), they showed that $\tilde{\varphi}$ has the following representation

$$\tilde{\varphi} = \mathbb{1}_{\{\tilde{z}\rho_T^* \tilde{L} H_\delta < \frac{d\bar{P}}{dP}\}} + \tilde{B}\mathbb{1}_{\{\tilde{z}\rho_T^* \tilde{L} H_\delta = \frac{d\bar{P}}{dP}\}}, \quad \text{P-a.s.} \qquad (6.23)$$

where $\tilde{B}$ is a $\mathcal{G}_T$-measurable random variable with values in $[0,1]$. In addition, $\tilde{\varphi}$, $\tilde{L}$, and $\tilde{z}$ satisfy the following conditions

$$E\left[\rho_T^* \tilde{L}\tilde{\varphi} H_\delta\right] = \tilde{u},$$

and if we introduce

$$\tilde{V}(\tilde{z}) := \inf_{L\in\mathcal{L}} E\left[\left(\frac{d\bar{P}}{dP} - \tilde{z}\rho_T^* L H_\delta\right)^+\right],$$

then $\tilde{V}(\tilde{z}) = E\left[\left(\frac{d\bar{P}}{dP} - \tilde{z}\rho_T^* \tilde{L} H_\delta\right)^+\right]$.

In the next lemma, we provide a more explicit description of $\tilde{L}$.

Lemma 6.3. *Consider the dual problem (6.22). For $\tilde{u} < U_0$, let $\tilde{V}(\tilde{z})$, $\tilde{z}$ and $\tilde{L}$ to be defined as above. Then*

(1) $\tilde{V}$ *vanishes, more precisely:*

$$\tilde{V}(\tilde{z}) = E\left[\left(\frac{d\bar{P}}{dP} - \tilde{z}\rho_T^* \tilde{L} H_\delta\right)^+\right] = 0, \qquad (6.24)$$

and

$$E^{\bar{P}}[\tilde{\varphi}] = V_*(\tilde{u}) = \tilde{u}\tilde{z}. \qquad (6.25)$$

(2) Moreover, we have:

$$\tilde{L}\prod_{i=1}^{n}\mathbb{1}_{\{\tau_i>T\}} = \left(\prod_{i=1}^{n}\mathbb{1}_{\{\tau_i>T\}}\right)\exp\left(\sum_{i=1}^{n}\int_0^T \mu_s^i ds\right). \quad (6.26)$$

Proof. (1) First, notice that by $l(0) = 0$ we can assume $\tilde{\varphi} = 1$ on the set $\{H_\delta = 0\}$. Now taking into account (6.23), we describe $\tilde{\varphi}$ on $\{H_\delta > 0\}$. From representation (6.23), we have $\tilde{\varphi} = 1$ on

$$A_1 := \left\{\tilde{z}\rho_T^*\tilde{L}H_\delta < \frac{d\bar{\mathrm{P}}}{d\mathrm{P}}\right\},$$

and $\tilde{\varphi} = \tilde{B}$ on

$$A_2 := \left\{\tilde{z}\rho_T^*\tilde{L}H_\delta = \frac{d\bar{\mathrm{P}}}{d\mathrm{P}}\right\} = \left\{\tilde{\alpha}\tilde{z}\rho_T^*\tilde{L} = l'\left((1-\tilde{\varphi})H_\delta\right)\right\}, \quad (6.27)$$

where $\tilde{\alpha} := \mathrm{E}\left[l'\left((1-\tilde{\varphi})H_\delta\right)H_\delta\right]$. Furthermore, it is clear that $\tilde{\varphi} = 0$ on

$$A_3 := (A_1 \cup A_2)^c.$$

Let us recall that we defined $\frac{d\bar{\mathrm{P}}}{d\mathrm{P}} = 0$ on $\tilde{\varphi} = 1$. On the other hand, we know all $\tilde{z}$, ρ_T^*, $\tilde{L}$ and H_δ are nonnegative. This implies that $A_1 = \emptyset$, and consequently $\tilde{V}(\tilde{z}) = 0$. Equation (6.25) is now obvious from (6.24) and (6.22).

(2) Without loss of generality, we suppose that $\delta = (\delta_1, \ldots, \delta_n) \in [0,1)^n$. In addition, let $j \in \{1, \ldots, n\}$ to be chosen such that $\delta_1, \ldots, \delta_j \in (0,1)$ and $\delta_i = 0$ for all $i = j+1, \ldots, n$ (up to a rearrangement of τ_i's). If $\delta_i = 0$ for all $i = 1, \ldots, n$ we take $j = 0$. For the case that $\delta \equiv 1$ see Remark 6.2.

Now, we split the proof into two cases:

(i) $\delta_i = 0$ for all $i = 1, \ldots, n$. We already proved that $A_1 = \emptyset$ and $\tilde{\varphi} = 0$ on A_3. In this case, we only need to investigate (6.27) on $\{H_\delta > 0\} = \{H > 0\} \cap \left(\bigcap_{i=1}^n \{\tau_i > T\}\right)$. Since for all $\kappa \in \mathcal{D}^n$ we have $\kappa > -1$, then

$$\rho_T^\kappa \prod_{i=1}^n \mathbb{1}_{\{\tau_i>T\}} \leq \bar{L} := \left(\prod_{i=1}^n \mathbb{1}_{\{\tau_i>T\}}\right)\exp\left(\sum_{i=1}^n\int_0^T \mu_s^i ds\right) \quad (6.28)$$

$$= \lim_{\substack{\bar{\kappa}\searrow -1 \\ \bar{\kappa}\text{ const}}} \rho_T^{\bar{\kappa}}\prod_{i=1}^n \mathbb{1}_{\{\tau_i>T\}} \in \mathcal{L}.$$

Defaultable securities and equity-linked life insurance contracts 147

In particular, this implies

$$\tilde{L} \prod_{i=1}^{n} \mathbb{1}_{\{\tau_i > T\}} \leqslant \bar{L}. \tag{6.29}$$

This inequality gives us

$$0 \leqslant \mathbb{E}\left[\left(\frac{d\bar{\mathbb{P}}}{d\mathbb{P}} - \tilde{z}\rho_T^* \bar{L} H_\delta\right)^+ \prod_{i=1}^n \mathbb{1}_{\{\tau_i > T\}}\right] \leqslant \tilde{V}(\tilde{z}). \tag{6.30}$$

Combining with part (1), $\bar{L}$ is, in fact, a solution to $\tilde{V}(\tilde{z})$.

(ii) For some $j \in \{1, \ldots, n\}$, $\delta_1, \ldots, \delta_j \in (0,1)$ and $\delta_i = 0$ for all $i = j+1, \ldots, n$. In this case, we have

$$\{H_\delta > 0\} = \{H > 0\} \cap \left(\bigcap_{i=j+1}^{n} \{\tau_i > T\}\right).$$

By (6.28) and (6.29), similar to (6.30) we can show that:

$$0 \leqslant \mathbb{E}\left[\left(\frac{d\bar{\mathbb{P}}}{d\mathbb{P}} - \tilde{z}\rho_T^* \bar{L} H_\delta\right)^+ \prod_{i=1}^n \mathbb{1}_{\{\tau_i > T\}}\right]$$

$$\leqslant \mathbb{E}\left[\left(\frac{d\bar{\mathbb{P}}}{d\mathbb{P}} - \tilde{z}\rho_T^* \tilde{L} H_\delta\right)^+ \prod_{i=1}^n \mathbb{1}_{\{\tau_i > T\}}\right] \leqslant \tilde{V}(\tilde{z}).$$

Using this and a similar argument as in case (i), (6.26) is proved. □

Remark 6.1. Let us point out that if the recovery rate $\delta \neq 0$, $\tilde{L}$ does not necessarily coincide with $\left(\prod_{i=1}^{n} \mathbb{1}_{\{\tau_i > T\}}\right) \exp\left(\sum_{i=1}^{n} \int_0^T \mu_s^i ds\right)$ on $\{H_\delta > 0\}$. In this case, inequality (6.29) does not hold on $\left(\bigcap_{i=1}^{j} \{\tau_i > T\}\right)^c \cap \left(\bigcap_{i=j+1}^{n} \{\tau_i > T\}\right) \cap \{H > 0\} \subset \{H_\delta > 0\}$. For instance, for $n = 1$ and $\delta \neq 0$, the family of $\mathcal{G}_T$-measurable random variables $\rho_T^\kappa \mathbb{1}_{\{\tau \leqslant T\}} = (1 + \kappa_\tau) \exp\left(-\int_0^\tau \kappa_s \mu_s ds\right) \mathbb{1}_{\{\tau \leqslant T\}}$, for $\kappa > -1$, does not possess an upper bound. However the modified option $\tilde{\varphi} H_\delta$, with $\tilde{\varphi}$ given by (6.23), still provides an implicit solution for the efficient hedging problem (6.15).

In the case of $\delta \equiv 0$, (6.26) fully describes $\tilde{L}$ on $\bigcap_{i=1}^{n} \{\tau_i > T\} \supseteq \{H_\delta > 0\}$. Considering this discussion, we will find an explicit representation for $\tilde{\varphi}$ when the recovery rate $\delta \equiv 0$.

Henceforth, we assume that the recovery rate $\delta_i = 0$ for all $i = 1, \ldots, n$. In particular, let us define

$$H_0 := H \prod_{i=1}^{n} \mathbb{1}_{\{\tau_i > T\}},$$

where H is a nonnegative $\mathcal{F}_T$-measurable random variable.

Theorem 6.1. *Under Assumption 6.3 on the loss function l, the optimal randomized test $\tilde{\varphi}$ is given by*

$$\tilde{\varphi} = \begin{cases} 1 - \left(I(\tilde{\lambda}\rho_T^*)/H_0\right) \wedge 1, & \{H_0 > 0\}; \\ 1, & \{H_0 = 0\}, \end{cases}$$

where the constant $\tilde{\lambda}$ can be determined by the constraint

$$\mathrm{E}^*[H - I(\tilde{\lambda}\rho_T^*) \wedge H] = \tilde{u}.$$

Moreover, $(\tilde{u}, \tilde{\pi})$ obtained from the optional decomposition of the modified claim $\tilde{\varphi}H_0$ gives us the the optimal strategy for the efficient hedging problem (6.15).

Proof. From (6.23) and Lemma 6.3 we can see that

$$\tilde{\varphi} = \tilde{B}\mathbb{1}_{\{\tilde{z}\rho_T^*\tilde{L}H_0 = \frac{d\tilde{\mathbb{P}}}{d\mathbb{P}}\}},$$

on $\{H_0 > 0\}$ and $\tilde{\varphi} = 1$ on $\{H_0 = 0\}$. Since $\delta \equiv 0$, recall that $\tilde{L} = \bar{L}$ by (6.28) and Remark 6.1. By some straightforward calculations, this becomes

$$\tilde{\varphi} = \begin{cases} 1 - I(\tilde{\lambda}\rho_T^*)/H_0, & \left\{\tilde{\varphi} = 1 - I(\tilde{\lambda}\rho_T^*)/H_0\right\} \cap \{H_0 > 0\}; \\ 0, & \left\{\tilde{\varphi} > 1 - I(\tilde{\lambda}\rho_T^*)/H_0\right\}; \\ 1, & \{H_0 = 0\}, \end{cases} \quad (6.31)$$

where $\tilde{\lambda} := \tilde{\alpha}\tilde{z}\exp\left(\sum_{i=1}^{n}\int_0^T \mu_s^i ds\right)$ is a constant. Expression (6.31) is still an implicit form for $\tilde{\varphi}$. To find an explicit representation, we define for $\lambda > 0$

$$\varphi_\lambda = \begin{cases} 1 - (I(\lambda\rho_T^*)/H_0) \wedge 1, & \{H_0 > 0\}; \\ 1, & \{H_0 = 0\}. \end{cases} \quad (6.32)$$

Because $\{\tau_i\}_{i=1,\ldots,n}$ and $\mathcal{F}_T$ are independent, we get

$$\mathrm{E}\left[\rho_T^*\tilde{L}\varphi_\lambda H_0\right] = \mathrm{E}^*\left[H - I(\lambda\rho_T^*) \wedge H\right].$$

By dominated convergence theorem, it is easy to see that $\mathrm{E}^*\left[H - I(\lambda\rho_T^*) \wedge H\right]$

decreases continuously from $E^*[H]$ to zero as λ increases from 0 to $+\infty$. Thus, for $\tilde{u} \in (0, E^*[H])$, there exists $\tilde{\lambda} > 0$ such that

$$E\left[\rho_T^* \tilde{L} \varphi_{\tilde{\lambda}} H_0\right] = \tilde{u}. \tag{6.33}$$

Let us consider $\varphi_{\tilde{\lambda}}$ defined by (6.32) and $\tilde{\lambda}$ chosen by (6.33). In the following, we show that $\varphi_{\tilde{\lambda}}$, in fact, satisfies (6.31).

On the set $\left\{\varphi_{\tilde{\lambda}} = 1 - I(\tilde{\lambda}\rho_T^*)/H_0\right\} \cap \{H_0 > 0\}$, it is clear that

$$I(\tilde{\lambda}\rho_T^*)/H_0 = 1 - \varphi_{\tilde{\lambda}} \leq 1.$$

This implies $\varphi_{\tilde{\lambda}} = 1 - \left(I(\tilde{\lambda}\rho_T^*)/H_0\right) \wedge 1 = 1 - I(\tilde{\lambda}\rho_T^*)/H_0$, the same as (6.31). Similarly, if $\varphi_{\tilde{\lambda}} > 1 - I(\tilde{\lambda}\rho_T^*)/H_0$ then one can see that

$$\left(I(\tilde{\lambda}\rho_T^*)/H_0\right) \wedge 1 = 1 - \varphi_{\tilde{\lambda}} < I(\tilde{\lambda}\rho_T^*)/H_0,$$

where the equality comes from the definition of $\varphi_{\tilde{\lambda}}$. This means $\left(I(\tilde{\lambda}\rho_T^*)/H_0\right) \wedge 1 = 1$, and again by the definition of $\varphi_{\tilde{\lambda}}$, it gives us $\varphi_{\tilde{\lambda}} = 0$. Finally, if we suppose that $\left\{\varphi_{\tilde{\lambda}} < 1 - I(\tilde{\lambda}\rho_T^*)/H_0\right\} \neq \emptyset$, we get the following contradiction

$$\left(I(\tilde{\lambda}\rho_T^*)/H_0\right) \wedge 1 > I(\tilde{\lambda}\rho_T^*)/H_0.$$

The last statement of the theorem is an immediate consequence of the existence of optional strategy for (6.16). □

Remark 6.2. $\delta \equiv 1$ implies that $H_\delta = H$ is default free in a complete market. In other words, for all τ_i's we have $\mathbb{1}_{\{\tau_i > T\}} \equiv 1$. By repeating the same arguments as above on $\{H > 0\}$ and $\{H = 0\}$ we get

$$\tilde{\varphi} = \begin{cases} 1 - \left(I(\tilde{\lambda}\rho_T^*)/H\right) \wedge 1 & ; \{H > 0\} \\ 1 & ; \{H = 0\} \end{cases}$$

such that $E^*[\tilde{\varphi} H] = \tilde{u}$.

Corollary 6.1. *For $\delta \equiv 0$ and given $\tilde{u} \in (0, E^*[H])$, the following conclusion holds:*

Consider $l(x) = \dfrac{x^p}{p}$ for some $p > 1$ and let $\tilde{\varphi}_p$ to be the corresponding $\tilde{\varphi}$ represented in Theorem 6.1. Then there exists $c > 0$ such that

$$(1 - \tilde{\varphi}_p) H_0 \mathbb{1}_{\{H_0 > 0\}} \longrightarrow (c \wedge H) \prod_{i=1}^{n} \mathbb{1}_{\{\tau_i > T\}} \mathbb{1}_{\{H > 0\}}$$

almost sure and also w.r.t $L^1(P^*)$-norm, as $p \longrightarrow +\infty$. The constant c is determined by $E^*[c \wedge H] = E^*[H] - \tilde{u}$.

Proof. Let us consider $\tilde{\lambda}_p$ as the corresponding $\tilde{\lambda}$ in Theorem 6.1. First of all we can show that for some $c > 0$

$$\lim_{p \to +\infty} \tilde{\lambda}_p^{\frac{1}{p-1}} = c.$$

Due to Theorem 6.1, we have

$$(1 - \tilde{\varphi}_p) H_0 \mathbb{1}_{\{H_0 > 0\}} = \left(\tilde{\lambda}_p^{\frac{1}{p-1}} (\rho_T^*)^{\frac{1}{p-1}} \wedge H \right) \prod_{i=1}^n \mathbb{1}_{\{\tau_i > T\}} \mathbb{1}_{\{H > 0\}},$$

and in addition

$$\tilde{u} = \mathrm{E}^* \left[\tilde{L} \tilde{\varphi}_p H_0 \mathbb{1}_{\{H_0 > 0\}} \right]$$

$$= \exp\left(\sum_{i=1}^n \int_0^T \mu_s^i ds \right) \mathrm{E}^* \left[\left(H - \tilde{\lambda}_p^{\frac{1}{p-1}} (\rho_T^*)^{\frac{1}{p-1}} \wedge H \right) \prod_{i=1}^n \mathbb{1}_{\{\tau_i > T\}} \right].$$

Since $\lim_{p \to +\infty} (\rho_T^*)^{\frac{1}{p-1}} = 1$, (P-a.s.), the above equations together with dominated convergence theorem prove the corollary. □

Now, let $ESR(\tilde{u})$ to be the minimum of the expectation of the shortfall risk for the default-free contingent claim H and initial capital $\tilde{u} < U_0 = \mathrm{E}^*[H]$, defined as (6.15).

Similarly, define $ESR^\tau(\tilde{u})$ as the minimum of the expectation of shortfall risk for H_0 and the available initial capital $\tilde{u}$. The next theorem provides a useful relation between $ESR(\tilde{u})$ and $ESR^\tau(\tilde{u})$.

Theorem 6.2. *Let $\tilde{u} \in (0, \mathrm{E}^*[H])$ to be given, then the following properties hold:*

(1) We have

$$ESR^\tau(\tilde{u}) = \exp\left(-\sum_{i=1}^n \int_0^T \mu_s^i ds \right) ESR(\tilde{u}). \tag{6.34}$$

(2) Suppose that $\hat{\pi} \in \mathcal{A}^{\mathbb{F}}(\tilde{u})$ is the optimal trading strategy that attains $ESR(\tilde{u})$ in the default-free market (6.1). Then the optimal trading strategy associated to $ESR^\tau(\tilde{u})$ is given by

$$(\tilde{\pi}_t)_{t \in [0,T]} = \left(\hat{\pi}_t \prod_{i=1}^n \mathbb{1}_{\{\tau_i \geq t\}} \right)_{t \in [0,T]} \in \mathcal{A}^{\mathbb{G}}(\tilde{u}). \tag{6.35}$$

Proof. (1) By the results of Theorem 6.1

$$ESR^\tau(\tilde{u}) = \mathrm{E}\left[l\left((1-\tilde{\varphi})H_0\right)\right]$$

$$= \mathrm{E}\left[l\left((I(\tilde{\lambda}\rho_T^*) \wedge H)\mathbb{1}_{\{H>0\}} \prod_{i=1}^n \mathbb{1}_{\{\tau_i>T\}}\right)\right] \quad (6.36)$$

$$= \mathrm{P}\left(\bigcap_{i=1}^n \{\tau_i > T\}\right) \mathrm{E}\left[l\left(I(\tilde{\lambda}\rho_T^*) \wedge H\right) \mathbb{1}_{\{H>0\}}\right].$$

The constant $\tilde{\lambda}$ can be determined from

$$\tilde{u} = \mathrm{E}^*\left[\left(H - I(\tilde{\lambda}\rho_T^*) \wedge H\right) \mathbb{1}_{\{H>0\}}\right], \quad (6.37)$$

keeping in mind this equation, Remark 6.2 implies

$$ESR(\tilde{u}) = \mathrm{E}\left[l\left(I(\tilde{\lambda}\rho_T^*) \wedge H\right) \mathbb{1}_{\{H>0\}}\right]. \quad (6.38)$$

Now, comparing (6.38) with (6.36) verifies equation (6.34).

(2) Since $\hat{\pi} \in \mathcal{A}^\mathbb{F}(\tilde{u})$ is a solution to $ESR(\tilde{u})$, and due to Remark 6.2

$$V_T^{\tilde{u},\hat{\pi}} = H - I(\tilde{\lambda}\rho_T^*) \wedge H,$$

where $\tilde{\lambda}$ satisfies (6.37). On the other hand, from Theorem 6.1 we know that the optional decomposition of $\tilde{\varphi}H_0$ gives us the the optimal solution corresponding to $ESR^\tau(\tilde{u})$. It is easy to see that

$$\tilde{\varphi}H_0 = \left(H - I(\tilde{\lambda}\rho_T^*) \wedge H\right) \prod_{i=1}^n \mathbb{1}_{\{\tau_i>T\}}$$

$$= V_T^{\tilde{u},\hat{\pi}} \prod_{i=1}^n \mathbb{1}_{\{\tau_i>T\}}.$$

Keeping in mind Assumptions 6.1 and 6.2, we apply the multidimensional Ito formula for $\left(V_t^{\tilde{u},\hat{\pi}} \prod_{i=1}^n \mathbb{1}_{\{\tau_i>t\}}\right)_{t\in[0,T]}$. Then the optional decomposition of $\tilde{\varphi}H_0$ is given as follows:

$$\tilde{\varphi}H_0 = \tilde{u} + \int_0^T \hat{\pi}_t \prod_{i=1}^n \mathbb{1}_{\{\tau_i \geq t\}} dS_t - V_{\tau_1}^{\tilde{u},\hat{\pi}} \mathbb{1}_{\{\tau_1 \leq T\}} \prod_{i=2}^n \mathbb{1}_{\{\tau_i>T\}}$$

$$- \sum_{i=1}^{n-1} \left(\int_0^{\tau_{i+1}} \hat{\pi}_t \prod_{j=1}^i \mathbb{1}_{\{\tau_j>t\}} dS_t\right) \mathbb{1}_{\{\tau_{i+1} \leq T\}} \prod_{k=i+2}^n \mathbb{1}_{\{\tau_k>T\}},$$

where we set $\prod_{k=i+2}^{n} \mathbb{1}_{\{\tau_k > T\}} \equiv 1$ for $i = n - 1$. We used the fact that the continuity of S allows us to write $\int_0^T \hat{\pi}_t \prod_{i=1}^{n} \mathbb{1}_{\{\tau_i > t\}} dS_t = \int_0^T \hat{\pi}_t \prod_{i=1}^{n} \mathbb{1}_{\{\tau_i \geq t\}} dS_t$. Using the decomposition of $\mathbb{G}$-predictable processes in terms of τ_is and the $\mathbb{F}$-predictable processes, we can claim that $\left(\hat{\pi}_t \prod_{i=1}^{n} \mathbb{1}_{\{\tau_i \geq t\}} \right)_{t \in [0,T]}$ is a $\mathbb{G}$-predictable process. In addition, we recall that $\mathbb{F}$-admissibility of $\hat{\pi}$ implies

$$\tilde{u} + \int_0^\theta \hat{\pi}_s dS_s \geq 0, \quad (\text{P-a.s.})$$

for all $\theta \in [0, T]$. Now, since $(\bigwedge_{i=1}^{n} \tau_i) \wedge t \in [0, T]$ we have

$$\tilde{u} + \int_0^t \hat{\pi}_s \prod_{i=1}^{n} \mathbb{1}_{\{\tau_i \geq s\}} dS_s = \tilde{u} + \int_0^{(\bigwedge_{i=1}^{n} \tau_i) \wedge t} \hat{\pi}_s dS_s \geq 0, \quad (\text{P-a.s.})$$

for all $t \in [0, T]$. This argument proves that, in fact, $\tilde{\pi} \in \mathcal{A}^{\mathbb{G}}(\tilde{u})$. $\square$

In fact, Theorem 6.2 reduces the efficient hedging problem in the defaultable market to the corresponding problem in the default-free market. The advantage of this result is to avoid the complication of working with the optional decomposition of $\tilde{\varphi} H_0$ in the enlarged filtration $\mathbb{G}$. By equations (6.34) and (6.35), for $\delta \equiv 0$ we only need to find the perfect hedging strategy of $H\left(1 - \left(I(\tilde{\lambda} \rho_T^*)/H\right) \wedge 1\right) \in \mathcal{F}_T$ to solve problem (6.15).

In the next lemma, we investigate the smoothness of the minimum of shortfall risk as a function of initial capital.

Lemma 6.4. *Let us consider $ESR^\tau : (0, \mathrm{E}^*[H]) \longrightarrow (\mathrm{E}[l(H_0)], 0)$ as a function of available initial capital $\tilde{u}$. Then $ESR^\tau \in C^1\left((0, \mathrm{E}^*[H])\right)$ and*

$$\frac{dESR^\tau}{du}(\tilde{u}) = -\tilde{z} \mathrm{E}\left[l'\left((1 - \tilde{\varphi}) H_0\right) H_0\right] \tag{6.39}$$

for all $\tilde{u} \in (0, \mathrm{E}^*[H])$.

Proof. Define $U : \Psi \longrightarrow \mathbb{R}$ as

$$U(\psi) := \mathrm{E}^*[\tilde{L} \psi H_0] \text{ for } \psi \in \Psi,$$

where $\Psi := \left\{ \psi \in L^1(\mathrm{P}) \mid \mathrm{E}^*[\tilde{L}\psi H_0] < +\infty \right\}$. Consider $\tilde{\varphi}$ defined as in Theorem 6.1, then by the formula for $\tilde{u}$ before Lemma 6.3

$$ESR^{\tau}(\tilde{u}) = ESR^{\tau}\left(U(\tilde{\varphi})\right).$$

To proceed, our idea is to exploit the Frechet derivative of ESR^{τ}, and the Gateaux derivative of U and $ESR^{\tau}oU$. By equation (6.34), it is clear that $ESR^{\tau} \in C^1\left((0, \mathrm{E}^*[H])\right)$ iff $ESR \in C^1\left((0, \mathrm{E}^*[H])\right)$.

It is not difficult to see that for $\delta \equiv 1$ (a complete market) $ESR \in C^1\left((0, \mathrm{E}^*[H])\right)$. It is also known that $ESR^{\tau} \in C^1\left((0, \mathrm{E}^*[H])\right)$ implies a Frechet differentiability of ESR^{τ}. Moreover, we can compute the Gateaux derivative of function U at $\tilde{\varphi}$ with the increment $\tilde{\varphi}$ as follows

$$DU(\tilde{\varphi}; \tilde{\varphi}) = \left. \frac{dU(\tilde{\varphi} + t\tilde{\varphi})}{dt} \right|_{t=0} = \mathrm{E}^*[\tilde{L}\tilde{\varphi} H_0] = \tilde{u}.$$

By the above arguments, the Frechet derivative of ESR^{τ} exists and U is Gateaux differentiable. Thus, we can apply the chain rule to evaluate the Gateaux derivative of $ESR^{\tau}oU$

$$D(ESR^{\tau}oU)(\tilde{\varphi}; \tilde{\varphi}) = DESR^{\tau}\left(U(\tilde{\varphi}); DU(\tilde{\varphi}; \tilde{\varphi})\right).$$

See, for instance, Kurdila and Zabarankin. On one hand, we have

$$D(ESR^{\tau}oU)(\tilde{\varphi}; \tilde{\varphi}) = \left. \frac{d\mathrm{E}\left[l\left((1 - \tilde{\varphi} - t\tilde{\varphi})H_0\right)\right]}{dt} \right|_{t=0}$$
$$= -\mathrm{E}\left[l'\left((1 - \tilde{\varphi})H_0\right)\tilde{\varphi}H_0\right] = -\tilde{\alpha}V_*(\tilde{u}) = -\tilde{\alpha}\tilde{u}\tilde{z},$$

where we used (6.25) and also recall that $\tilde{\alpha} = \mathrm{E}[l'\left((1 - \tilde{\varphi})H_0\right)H_0]$. On the other hand

$$DESR^{\tau}\left(U(\tilde{\varphi}); DU(\tilde{\varphi}; \tilde{\varphi})\right) = DESR^{\tau}(\tilde{u}; \tilde{u})$$
$$= \left. \frac{dESR^{\tau}(\tilde{u} + t\tilde{u})}{dt} \right|_{t=0} = \tilde{u}\frac{dESR^{\tau}}{du}(\tilde{u}).$$

Finally, combining all these equalities together, we obtain (6.39)

$$\frac{dESR^{\tau}}{du}(\tilde{u}) = -\tilde{\alpha}\tilde{z} < 0$$

for all $\tilde{u} \in (0, \mathrm{E}^*[H])$. □

With the help of the above lemma, we can provide more qualitative features of $\tilde{\varphi}$ and $\tilde{z}$ corresponding to maximization problem (6.17) and its dual problem (6.22), for $\delta \equiv 0$.

Lemma 6.5. *Let us consider $\tilde{z}$, $\tilde{\lambda}$, $\tilde{\alpha}$, and $\tilde{\varphi}$ (defined above) as functions of $\tilde{u} \in (0, \mathrm{E}^*[H])$. Assume that $\{\tilde{u}_m\}_{m \geqslant 0} \subset (0, \mathrm{E}^*[H])$ and $\tilde{u}_m \longrightarrow \tilde{u}_0$ as $m \longrightarrow \infty$. Then $\lim\limits_{m \to +\infty} \tilde{\lambda}(\tilde{u}_m) = \tilde{\lambda}(\tilde{u}_0)$, moreover*

$$\tilde{\varphi}(\tilde{u}_m) \longrightarrow \tilde{\varphi}(\tilde{u}_0) \quad \text{P-a.s. and w.r.t } L^1(\mathrm{P})\text{-norm} \tag{6.40}$$

as $m \longrightarrow \infty$. In particular, we have $\lim\limits_{m \to +\infty} \tilde{z}(\tilde{u}_m) = \tilde{z}(\tilde{u}_0)$.

Proof. By $\tilde{\lambda}(u) = \tilde{\alpha}(u)\tilde{z}(u) \exp\left(\sum\limits_{i=1}^{n} \int_0^T \mu_s^i \, ds\right)$ and $\tilde{\alpha}(u)\tilde{z}(u) = -\dfrac{dESR^\tau}{du}(u) \in C((0, \mathrm{E}^*[H]))$, it is clear that

$$\lim_{m \to +\infty} \tilde{\lambda}(\tilde{u}_m) = \tilde{\lambda}(\tilde{u}_0).$$

Using the representation of $\tilde{\varphi}$ in Theorem 6.1, a continuity of I and the dominated convergence theorem, we can prove (6.40).

Since $\mathrm{E}^*\left[\tilde{L}\tilde{\varphi}(\tilde{u}_m) H_0\right] = \tilde{u}_m \in (0, \mathrm{E}^*[H])$ for all $m \geqslant 0$, it is easy to see that $\tilde{\varphi}(\tilde{u}_m) \neq 1$ P-a.s and as a result $\tilde{\alpha}(\tilde{u}_m) \neq 0$. Therefore, the continuity of $\tilde{z}(.)$ on $(0, \mathrm{E}^*[H])$ can be deduced from the same property for $\tilde{\alpha}(.)$ and $\tilde{\lambda}(.)$. The following inequality, (6.40) and dominated convergence theorem together establish the continuity of $\tilde{\alpha}(.)$:

$$0 \leqslant \mathrm{E}\left[l'\left((1 - \tilde{\varphi}(\tilde{u}_m))H_0\right) H_0\right] \leqslant \mathrm{E}\left[l'\left(I(\tilde{\lambda}(\tilde{u}_m)\rho_T^*)\right) H\right] = \tilde{\lambda}(\tilde{u}_m)\mathrm{E}^*[H].$$

□

To demonstrate our results, we consider the power function $l(x) = \dfrac{x^p}{p}$ for some $p > 0$. In this case, problem (6.15) turns into a problem of minimizing the lower partial moments with the random target H_0.

Example 6.1. Assume $n = 1$, $\delta \equiv 0$, and $H = (S_T - K)^+$ for some $K > 0$ as the strike price of the call option H. Hence,

$$H_0 = H \mathbb{1}_{\{\tau > T\}}.$$

Working in the framework of the Black–Scholes model with constant parameters σ and $m > 0$, we get

$$\frac{d\mathrm{P}^*}{d\mathrm{P}} := \rho_T^* = \exp\left(-\frac{m}{\sigma}W_T - \frac{1}{2}\left(\frac{m}{\sigma}\right)^2 T\right)$$

$$= S_0^{\frac{m}{\sigma^2}} \exp\left(\frac{m^2}{2\sigma^2}T - \frac{1}{2}mT\right) S_T^{-\frac{m}{\sigma^2}}$$

for $t \in [0, T]$. Clearly, by Girsanov's theorem, $(W_t^*)_{0 \leqslant t \leqslant T} := \left(W_t + \frac{m}{\sigma}t\right)_{0 \leqslant t \leqslant T}$ is an $(\mathbb{F}, \mathrm{P}^*)$ standard Brownian motion.

Now, by our results, problem (6.15) can be solved in two ways:

(1) Directly, using our result for defaultable markets (i.e., Theorem 6.1). In this case, we need to find the optional decomposition of $\tilde{\varphi}H_0$ in the enlarged filtration $\mathbb{G}$. However, this method demands some tedious calculations and finally gives us a complicated hedging strategy.

Suppose $\tilde{u} < \mathrm{E}^*[(S_T - K)^+]$, then Theorem 6.1 implies

$$\tilde{\varphi}_p H_0 = \begin{cases} H - (\tilde{\lambda}\rho_T^*)^{\frac{1}{p-1}} \wedge H, & \{H > 0\} \cap \{\tau > T\}; \\ 0, & \{H = 0\} \cup \{\tau \leqslant T\}. \end{cases}$$

We know that $(\tilde{u}, \tilde{\pi})$ obtained from the optional decomposition of the modified claim $\tilde{\varphi}_p H_0$ solves the efficient hedging problem (6.15). Similar to Section 6.1, for $t \in [0, T]$ define

$$\tilde{X}_t := \operatorname*{ess\,sup}_{\kappa \in \mathcal{D}} \mathrm{E}^{Q^\kappa}\left[\tilde{\varphi}_p H_0 | \mathcal{G}_t\right].$$

For $\kappa \in \mathcal{D}$ and $t \in [0, T]$, first we simplify the underlying conditional expectation:

$$\mathrm{E}^{Q^\kappa}\left[\tilde{\varphi}_p H_0 | \mathcal{G}_t\right] = \frac{1}{\rho_t^\kappa \rho_t^*} \exp\left(-\int_0^t \kappa_s \mu_s ds\right)$$

$$\times \mathrm{E}\left[\rho_T^* \exp\left(-\int_t^T \kappa_s \mu_s ds\right)\left(H - (\tilde{\lambda}\rho_T^*)^{\frac{1}{p-1}} \wedge H\right) \mathbb{1}_{\{\tau > T\}} \Big| \mathcal{G}_t\right].$$

Let $\tilde{H} := H - (\tilde{\lambda}\rho_T^*)^{\frac{1}{p-1}} \wedge H$. Then some calculations give us

$$\mathrm{E}\left[\rho_T^* \exp\left(-\int_t^T \kappa_s \mu_s ds\right) \tilde{H} \mathbb{1}_{\{\tau > T\}} \Big| \mathcal{G}_t\right]$$

$$= \mathbb{1}_{\{\tau > t\}} \mathrm{E}\left[\rho_T^* \tilde{H} \exp\left(\int_0^t \mu_s ds - \int_t^T \kappa_s \mu_s ds\right) \mathbb{1}_{\{\tau > T\}} \Big| \mathcal{F}_t\right].$$

Therefore, by the above

$$\mathrm{E}^{Q^\kappa}\left[\tilde{\varphi}_p H_0 | \mathcal{G}_t\right]$$

$$= \frac{1}{\rho_t^*} \mathbb{1}_{\{\tau > t\}} \mathrm{E}\left[\rho_T^* \tilde{H} \exp\left(\int_0^t \mu_s ds - \int_t^T \kappa_s \mu_s ds\right) \mathbb{1}_{\{\tau > T\}} \Big| \mathcal{F}_t\right]. \quad (6.41)$$

Now, if κ is constant in (6.41) and $\kappa \searrow -1$ then by Fatou's lemma and the definition of $\tilde{X}_t$ we get

$$\tilde{X}_t \geq \frac{1}{\rho_t^*} \mathbb{1}_{\{\tau>t\}} E\left[\rho_T^* \tilde{H} \exp\left(\int_0^T \mu_s ds\right) \mathbb{1}_{\{\tau>T\}} \Big| \mathcal{F}_t\right]$$

$$= \frac{1}{\rho_t^*} \mathbb{1}_{\{\tau>t\}} \exp\left(\int_0^T \mu_s ds\right) E\left[\rho_T^* \tilde{H} E\left[\mathbb{1}_{\{\tau>T\}} \Big| \mathcal{F}_T\right] \Big| \mathcal{F}_t\right] \quad (6.42)$$

$$= \mathbb{1}_{\{\tau>t\}} E^*\left[\tilde{H} \Big| \mathcal{F}_t\right].$$

On the other hand, due to (6.41) and $\kappa > -1 \quad ds \times dP$-a.e. it can be seen that

$$\tilde{X}_t \leq \frac{1}{\rho_t^*} \mathbb{1}_{\{\tau>t\}} E\left[\rho_T^* \tilde{H} \exp\left(\int_0^T \mu_s ds\right) \mathbb{1}_{\{\tau>T\}} \Big| \mathcal{F}_t\right] \quad (6.43)$$

$$= \mathbb{1}_{\{\tau>t\}} E^*\left[\tilde{H} \Big| \mathcal{F}_t\right].$$

The inequalities (6.42) and (6.43) show that

$$\tilde{X}_t = \mathbb{1}_{\{\tau>t\}} E^*\left[\tilde{H} \Big| \mathcal{F}_t\right]$$
$$= E^*\left[\tilde{H} \Big| \mathcal{F}_t\right] - \mathbb{1}_{\{\tau\leq t\}} E^*\left[\tilde{H} \Big| \mathcal{F}_t\right] \quad (6.44)$$

for $t \in [0, T]$.

By martingale representation theorem for Brownian filtrations, we have

$$E^*\left[\tilde{H} \Big| \mathcal{F}_t\right] = E^*[\tilde{H}] + \int_0^t \pi'_u dS_u$$

for some $\mathbb{F}$-predictable process π'.

Applying Ito's formula on the second term of (6.44) (in the second equality) and using the above representation along with the continuity of the process S, we get

$$\tilde{X}_t = \tilde{u} + \int_0^t \pi'_u \mathbb{1}_{\{\tau \geq u\}} dS_u - \left(\tilde{u} + \int_0^\tau \pi'_u dS_u\right) N_t \quad (6.45)$$

where $N_t := \mathbb{1}_{\{\tau \leq t\}}$. Furthermore, notice that similar to (6.37)

$$\tilde{u} = E^*\left[H - (\tilde{\lambda}\rho_T^*)^{\frac{1}{p-1}} \wedge H\right].$$

We can interpret the optional decomposition (6.45) as follows: Starting with $\tilde{u}$ as the initial capital, if we hold $\pi'_t \mathbb{1}_{\{\tau \geq t\}}$ number of shares of the stock at time $t \in [0,T]$, and withdraw the amount $\left(\tilde{u} + \int_0^\tau \pi'_u dS_u\right) N_t$ then we can guarantee to generate $\tilde{\varphi}_p H_0$ at time $t = T$.

(2) In contrast to the above, we can apply Theorem 6.2 and Remark 6.2. In other words, instead of solving the efficient hedging problem in the defaultable market, we first solve our problem in the complete market. Then, we apply (6.34) and (6.35) to determine the minimum of shortfall risk and the optimal strategy in the defaultable market.

Keeping in mind the second approach, let us fix $r \equiv 0$, $m = 0.02$, $\sigma = 0.2$, $S_0 = 1$, $K = 0.8$, and $T = 15$ (years), thus $U_0 = \mathrm{E}^*[(S_T - K)^+] = 0.3819$. In addition, assume $\mu \equiv 0.01$ which implies $\mathrm{P}(\tau > T) = 0.8607$ (a probability of $\mathrm{P}(\tau \leq T) = 0.1393$ default before the maturity time $T = 15$).

Consider $l(x) = \dfrac{x^2}{2}$ and the available initial capital $\tilde{u} = 0.17 < U_0$ to hedge H_0. Applying Remark 6.2, we have

$$ESR(\tilde{u}) = 0.0971.$$

In the Section 6.3, for an analogous claim $H = (S_T - K)^+ + K$, we provide the details for how to compute $ESR(\tilde{u}) = \frac{1}{p}\mathrm{E}\left[((1 - \tilde{\varphi}(\tilde{u}))H)^p\right]$ and the associated optimal trading strategy $\hat{\pi}$.

By Theorem 6.2 and above, starting with $\tilde{u} = 0.17$, the minimum of the expectation of shortfall risk weighted by l for H_0 becomes

$$ESR^\tau(\tilde{u}) = \mathrm{P}(\tau > T)ESR(\tilde{u}) = 0.8607 \times 0.0971 = 0.0836.$$

For some fixed values of initial capital, Table 6.1 presents the associated minimum shortfall risk versus $\tilde{u}$. For a given $\tilde{u}$, since $H_0 \leq H$, as it is expected $ESR^\tau(\tilde{u})$ is less than $ESR(\tilde{u})$.

Initial capital	$ESR^\tau(\tilde{u})$	$ESR(\tilde{u})$
$\tilde{u}$	$\delta = 0$	$\delta = 1$
$0.32	0.0051	0.0059
$0.27	0.0183	0.0213
$0.22	0.0429	0.0498
$0.17	0.0836	0.0971

TABLE 6.1: $ESR^\tau(\tilde{u})$ vs. $ESR(\tilde{u})$ for a defaultable call option.

6.3 Application to equity-linked life insurance contracts: Brennan–Schwartz approach vs. the superhedging

We study equity-linked life insurance contracts in the framework of Section 6.2 here. Although there are different types of equity-linked life insurance contracts, we will concentrate on a pure endowment contract:

$$H\mathbb{1}_{\{T(x)>T\}} \tag{6.46}$$

where H is a nonnegative $\mathcal{F}_T$-measurable random variable and $T(x)$ is a positive random variable defined on the probability space $(\Omega, \mathcal{G}, \mathrm{P})$. In fact, H is a future payment at time $t = T$ the size of which depends on the evolution of the risky asset S during the contract period $[0, T]$, and $T(x)$ represents the remaining lifetime (or the future lifetime) of a client who is currently at age x. The quantity as usual

$$_T p_x := \mathrm{P}\left(T(x) > T\right)$$

is called the survival probability of the client. Using "life tables" (see for instance, Bowers et al.) we can find $_T p_x$ of each client for our pricing and hedging purposes. Clearly, $_T p_x$ depends on some factors such as age, race, sex, ets., We do not touch the mortality modeling, while an appropriate stochastic mortality modeling can bring reasonable advantages in such pricings. For a pure endowment contract (6.46), if the insured is still alive at maturity of the contract, the payment is H, otherwise zero.

Similarly, we can define a defaultable (pure endowment) equity-linked life insurance contract with recovery rate δ as a contract with the following payoff function

$$H_\delta\left(\tau, T(x)\right) := \left(H\mathbb{1}_{\{\tau > T\}} + \delta H\mathbb{1}_{\{\tau \leqslant T\}}\right)\mathbb{1}_{\{T(x)>T\}} \tag{6.47}$$

where τ is a default time for the insurance company. Therefore, to receive the payment H, the client must be alive at time T and also the insurance company should not default up to this time. In the following, to provide explicit solutions (by applying Theorem 6.1), we let $\delta \equiv 0$. In this case, (6.47) is denoted by $H_0\left(\tau, T(x)\right)$.

Assumption 6.4. *We postulate that S, $T(x)$, and τ are mutually independent.*

The three elements of our model, S, $T(x)$ and τ, generate two types of risks. There is an uncertainty associated to the asset price and the default time. This risk depends on the behavior of the financial market, and it is known as financial/credit risk from financial literature. Another source of risk is the so-called mortality risk from insurance terminology, and it is the risk caused by the mortality time of the client, $T(x)$, which is independent of the financial market. There are different approaches to hedge and price the contingent claim $H_0(\tau, T(x))$; we focus on the superhedging approach and Brennan–Schwartz approach to deal with these two sources of risk, respectively.

By the Brennan–Schwartz approach, the size of the life insurance contracts is considered to be large enough to use the strong law of large numbers. In other words, if the insurer sells the insurance contract (6.47) to N clients, then we have

$$\sum_{i=1}^{N} \mathbb{1}_{\{T_i(x) > T\}} \approx N {}_T p_x.$$

This means that by applying the strong law of large numbers, the mortality risk is managed (diversified) by the size of the contracts. Hence, hedging $H_0(\tau, T(x))$ reduces to hedging the modified claim ${}_T p_x H \mathbb{1}_{\{\tau > T\}}$.

Keeping in mind that $\mathcal{G}_t = \mathcal{F}_t \vee \mathcal{H}_t$ and $\mathcal{H}_t = \sigma(\tau \wedge t)$ for $t \in [0, T]$. To hedge the credit risk associated to $H_0(\tau, T(x))$, we apply superhedging techniques for $H \mathbb{1}_{\{\tau > T\}}$ in the incomplete market (B, S, τ) equipped with the filtration $\mathbb{G} = (\mathcal{G}_t)_{0 \leq t \leq T}$. In fact, for a single contract $H_0(\tau, T(x))$ the insurance company should superhedge ${}_T p_x$ short positions of $H \mathbb{1}_{\{\tau > T\}}$ in the defaultable market.

As a particular case, we study equity-linked life insurance contracts with constant guarantee K, i.e.,

$$H = \max(S_T, K) = (S_T - K)^+ + K$$

in (6.47). Using the Brennan–Schwartz argument and Lemma 6.1, we consider the following amount as the premium for the insurance contract (6.47)

$$\tilde{u} := {}_T p_x U_0 = {}_T p_x \mathrm{E}^* [\max(S_T, K)]$$
$$= {}_T p_x \mathrm{E}^* [(S_T - K)^+] + {}_T p_x K$$

obviously, $\tilde{u} < U_0 = \mathrm{E}^* [\max(S_T, K)]$.

The same as Example 6.1, we take $l(x) = \dfrac{x^p}{p}$ for some $p > 1$. Starting with the premium $\tilde{u} = {}_T p_x U_0$ as the initial capital, we want to solve the efficient hedging problem (6.15) for the insurance contract $H_0(\tau, T(x))$. Let $\hat{\varphi}_p$ be the optimal solution corresponding to the initial capital $\tilde{u}$ and the problem introduced in Remark 6.2. Then Theorems 6.1 and 6.2 show that the perfect hedging of the modified claim $\hat{\varphi}_p \cdot \max(S_T, K)$ solves our efficient hedging problem. We follow Sections 6.1 and 6.2 to find the explicit solution. By Remark 6.2:

$$\hat{\varphi}_p \cdot \max(S_T, K) = \max(S_T, K) - \left(c^{\frac{1}{p-1}} S_T^{\frac{-\beta}{p-1}} \right) \wedge \max(S_T, K), \quad (6.48)$$

where $\beta := \dfrac{m}{\sigma^2}$, and the constant c comes from the constants involving Theorem 6.1 and formula ρ_T^* in Example 6.1. By $\tilde{u} = \mathrm{E}^* [\hat{\varphi}_p \cdot \max(S_T, K)]$, depending on the value of $\tilde{u}$, the decreasing convex function $c^{\frac{1}{p-1}} s^{\frac{-\beta}{p-1}}$ intersects with $\max(s, K)$ at $s = K_1 < K$ or $s = K_2 \geq K$. More precisely, we have:

(i) If $\tilde{u} > \mathrm{E}^*\left[\left(S_T - K(\frac{S_T}{K})^{\frac{\beta}{1-p}}\right) \mathbb{1}_{\{S_T \geq K\}}\right]$ then (6.48) becomes

$$= K\mathbb{1}_{\{K_1 \leq S_T < K\}} + S_T \mathbb{1}_{\{S_T \geq K\}} - K\left(\frac{S_T}{K_1}\right)^{\frac{\beta}{1-p}} \mathbb{1}_{\{S_T \geq K_1\}}. \quad (6.49)$$

(ii) If $\tilde{u} \leq \mathrm{E}^*\left[\left(S_T - K\left(\frac{S_T}{K}\right)^{\frac{\beta}{1-p}}\right) \mathbb{1}_{\{S_T \geq K\}}\right]$ then, in this case, (6.48) is equal to

$$= S_T \mathbb{1}_{\{S_T \geq K_2\}} - K_2 \left(\frac{S_T}{K_2}\right)^{\frac{\beta}{1-p}} \mathbb{1}_{\{S_T \geq K_2\}}.$$

Now, applying the results of Section 6.2, we provide an analytic expression for the minimum value of shortfall risk. Additionally, the optimal strategy is derived by using the replication principle in complete markets. We only provide the details for the first case, (6.49); the corresponding results for the second case can be obtained by some straightforward modifications. Let us define

$$V_t := \mathrm{E}^*\left[\hat{\varphi}_p \cdot \max\left(S_t \exp\left[\sigma(W_T^* - W_t^*) - \frac{1}{2}\sigma^2(T-t)\right], K\right) \Big| \mathcal{F}_t\right] = F_p(t, S_t)$$

for $t \in [0, T]$. In the case of (6.49), the Markov property and lognormal distribution of S_t imply that

$$F_p(t, s) = K\Phi\left(d_-(t, s, K_1)\right) - K\Phi\left(d_-(t, s, K)\right) + s\Phi\left(d_+(t, s, K)\right)$$

$$-K\left(\frac{s}{K_1}\right)^{\frac{\beta}{1-p}} \exp\left[\frac{m(T-t)}{2(p-1)}\left(\frac{\beta}{p-1} + 1\right)\right] \Phi\left(d_-(t, s, K_1) + \frac{m\sqrt{T-t}}{\sigma(1-p)}\right),$$

where Φ is the standard normal distribution function and

$$d_\pm(t, s, K) = \frac{\ln s - \ln K}{\sigma\sqrt{T-t}} \pm \frac{1}{2}\sigma\sqrt{T-t}.$$

The constant K_1, and a priori c, can be determined from

$$\tilde{u} = \mathrm{E}^*\left[\hat{\varphi}_p \cdot \max(S_T, K)\right] = F_p(0, S_0).$$

After finding K_1, by Theorem 6.2 the minimum shortfall risk can be calculated as follows

$$ESR^\tau(\tilde{u}) = \mathrm{P}(\tau > T) ESR(\tilde{u}) = \frac{1}{p}\mathrm{P}(\tau > T) \mathrm{E}\left[\left((1 - \hat{\varphi}_p(\tilde{u}))H\right)^p\right]$$

$$= \frac{K^p}{p}\mathrm{P}(\tau > T) \left\{1 - \Phi\left(d_-(0, S_0, K_1) + \frac{m\sqrt{T}}{\sigma}\right)\right.$$

$$+ \left(\frac{S_0}{K_1}\right)^{\frac{p\beta}{1-p}} \exp\left[\frac{p\beta T}{p-1}\left(\frac{1}{2}\sigma^2\left(\frac{p\beta}{p-1} + 1\right) - m\right)\right] \quad (6.50)$$

$$\left. \times \Phi\left(d_-(0, S_0, K_1) + \frac{m\sqrt{T}}{\sigma} + \frac{mp\sqrt{T}}{\sigma(1-p)}\right)\right\}.$$

Moreover, the optimal strategy corresponding to $ESR(\tilde{u})$ is given by

$$\hat{\pi}_t = \frac{\partial}{\partial s} F_p(t,s)|_{s=S_t}$$

$$= \left\{ \frac{K}{s\sigma\sqrt{2\pi(T-t)}} \left[\exp\left(-\frac{d_-^2(t,s,K_1)}{2}\right) - \exp\left(-\frac{d_-^2(t,s,K)}{2}\right) \right. \right.$$

$$\left. + \frac{s}{K} \exp\left(-\frac{d_+^2(t,s,K)}{2}\right) \right] + \Phi(d_+(t,s,K))$$

$$- \frac{K}{s} \left(\frac{s}{K_1}\right)^{\frac{\beta}{1-p}} \exp\left[\frac{m(T-t)}{2(1-p)}\left(\frac{\beta}{p-1}+1\right)\right] \quad (6.51)$$

$$\times \left[\frac{\beta}{1-p} \Phi\left(d_-(t,s,K_1) + \frac{m\sqrt{T-t}}{\sigma(1-p)}\right) \right.$$

$$\left. \left. + \frac{1}{\sigma\sqrt{2\pi(T-t)}} \exp\left(-\frac{\left(d_-(t,s,K_1) + \frac{m\sqrt{T-t}}{\sigma(1-p)}\right)^2}{2}\right) \right] \right\} \Bigg|_{s=S_t}.$$

Therefore, by (6.35) and (6.51), the trading strategy $(\hat{\pi}_t \mathbb{1}_{\{\tau \geq t\}})_{t \in [0,T]}$ is a solution to $ESR^\tau(\tilde{u})$.

As an alternative to the above discussion (Brennan–Schwartz method), it is possible to treat both τ and $T(x)$ as independent default times. In this case, we can apply superhedging techniques developed before and where the filtration is enlarged by both τ and $T(x)$, i.e.,

$$\mathcal{G}_t = \mathcal{F}_t \vee \mathcal{H}_t.$$

Here $\mathcal{H}_t = \sigma(\tau \wedge t) \vee \sigma(T(x) \wedge t)$ for $t \in [0,T]$. We know that there exists a nonnegative function $\hat{\mu}$ such that for all $t \geq 0$

$$_tp_x := P(T(x) > t) = \exp\left(-\int_0^t \hat{\mu}(x+s)ds\right).$$

$\hat{\mu}$ is known as the force of mortality or the hazard rate function.

Then considering $\tau_1 = \tau$, $\tau_2 = T(x)$, and $\delta_1 = \delta_2 = 0$ in (6.5), we can use Theorem 6.2, relation (6.34), to get

$$ESR^{\tau,T(x)}(\tilde{u}) = {}_Tp_x P(\tau > T) ESR(\tilde{u}). \quad (6.52)$$

Notice that $ESR(\tilde{u})$ can be computed as presented before in the Brennan–Schwartz approach. Relation (6.52) means that if we add the information regarding the survival of clients at each $t \in [0,T]$ to the filtration $\mathbb{G}$, then the minimum of shortfall risk is reduced by the ratio of $_Tp_x$.

Example 6.2. Let us consider $l(X) = \dfrac{X^2}{2}$ and the parameters of our model

162   EQUITY-LINKED LIFE INSURANCE   Partial Hedging Methods

same as Example 6.1. In this example, we suppose that the client is at age $x = 30$. For $T = 15$ and $x = 30$, using the life tables in Bowers et al., the client will survive to the maturity time of the contract with the probability of $_Tp_x = 0.949$. Moreover, it is easy to see that $U_0 = E^*[\max(S_T, K)] = 1.182$.

Taking the Brennan–Schwartz approach into consideration and starting with the premium $\tilde{u} = {}_Tp_x U_0 = 1.122$ as the initial capital, we can employ (6.50) to see that

$$ESR^\tau(\tilde{u}) = 0.0014$$

for $\delta \equiv 0$, and $ESR(\tilde{u}) = 0.0016$ for $\delta = 1$.

In the case of the superhedging approach, with the same $\tilde{u} = 1.122$, we apply (6.52) this time to obtain

$$ESR^{\tau, T(x)}(\tilde{u}) - 0.0013$$

for $\delta_1 = \delta_2 = 0$, and $ESR^{T(x)}(\tilde{u}) = 0.0015$ where $\delta_1 = 1$ and $\delta_2 = 0$.

In Table 6.2, we compare Brennan–Schwartz and superhedging methods for some given values of $\tilde{u}$. Similar to Table 6.1, from the insurer's point of view, $ESR^\tau(.)$ is still a decreasing function of the initial capital, and for a fixed $\tilde{u}$ it decreases with a higher possibility of the default event.

Obviously, using the superhedging approach generates smaller values for $ESR^\tau(\tilde{u})$, and this is consistent with our intuition. In the case of the Brennan–Schwartz method, we eliminate the mortality risk of the clients by the constant number $_Tp_x$, but in the superhedging method, information that is more accurate is available. In the latter case, with the enlargement of the filtrations and adding the new source of randomness to our model, we can provide a better approximation of the risk.

Initial capital	Brennan–Schwartz approach		Superhedging approach	
	$ESR^\tau(\tilde{u})$	$ESR(\tilde{u})$	$ESR^{\tau,T(x)}(\tilde{u})$	$ESR^{T(x)}(\tilde{u})$
$\tilde{u}$	$\delta = 0$	$\delta = 1$	$\delta_1 = \delta_2 = 0$	$\delta_1 = 1, \delta_2 = 0$
$1.122	0.0014	0.0016	0.0013	0.0015
$1.066	0.0050	0.0058	0.0048	0.0055
$1.010	0.0110	0.0128	0.0105	0.0121
$0.954	0.0193	0.0225	0.0184	0.0213

TABLE 6.2: A comparison of the minimum shortfall risk for equity-linked life insurance contracts with guarantee: the Brennan–Schwartz approach vs. the superhedging approach.

Chapter 7

Equity-linked life insurance contracts and Bermudan options

7.1 GMDB life insurance contracts as Bermudan options 163
7.2 Quantile hedging of GMDB life insurance contracts 170
7.3 Numerical illustrations 181

7.1 GMDB life insurance contracts as Bermudan options

We study guaranteed minimum death benefit (GMDB) life insurance contracts where the benefit is paid upon the insured's death over the life of the contract. The payoff process of a GMDB contract with a constant guarantee is given by

$$U_t := \max(K, S_t), \quad \text{for} \quad t \in R := \{1, 2, ..., T\}, \tag{7.1}$$

where $K > 0$ is the constant amount of the guarantee and $(S_t)_{t \in [0,T]}$ is the price process of the underlying asset. The finite set R is a suitable subset of $[0, T]$, for instance months.

Using the techniques of enlargement of filtrations, we construct a framework which allows us to view the GMDB contract as an American option with the predetermined finite exercise dates R. These types of American options are known as Bermudan options. In this setting, one can also compare the GMDB contract (7.1) with the option based portfolio insurance (OBPI). Let us denote the filtration generated by S by $\mathcal{F}$ and the filtration generated by the client's lifetime by $\mathcal{H}$. Later we provide more precise definitions for the filtrations $\mathcal{F}$ and $\mathcal{G}$. To make the exercise date of the GMDB contract a stopping time, we progressively enlarge $\mathcal{F}$ with $\mathcal{H}$ and denote the enlarged filtration by $\mathcal{G}$.

Assume that $\tilde{v}_0 > 0$ and a $\mathcal{G}$-predictable S-integrable process $(\pi_t)_{t \in [0,T]}$ are given. Define the corresponding value process as follows:

$$V_t^{\tilde{v}_0, \pi} := \tilde{v}_0 + \int_0^t \pi_s dS_s, \quad \text{P-a.s.,} \quad \text{for all} \quad t \in [0, T].$$

We represent the set of all processes π satisfying

$$V_t^{\tilde{v}_0, \pi} \geq 0 \text{ P-a.s. for all } t \in [0, T]$$

by $\mathcal{A}^{\mathcal{G}}(\tilde{v}_0)$.

Using the independency assumption, as described above, we provide an explicit form for the Radon–Nikodym density of the martingale probability measures of S with respect to the new filtration $\mathcal{G}$.

We develop the superhedging method for Bermudan options, which approach, is adapted to price the embedded Bermudan option in the GMDB contract. From the independency and the Radon–Nikodym density representation, we show that the superhedging value process of the GMDB contract is equal to the perfect hedging for the European option $\max(K, S_T)$. In addition, with the separate account design of the GMDB, the actual liability of the insurance company becomes $(K - S_T)^+$, i.e., the shortfall in the case where the guarantee K matures in-the-money.

The main aim is to solve the quantile hedging problem for the Bermudan option $(U_t)_{t \in R}$. More precisely, the max-min problem (7.2) is solved and the optimal trading strategy that achieves the maximal value is determined. Let $\mathcal{S}_{0,T}(R)$ be the set of all $\mathcal{G}$-stopping with values in R, then we investigate the following problem:

$$\sup_{\pi \in \mathcal{A}^{\mathbb{G}}(\tilde{v}_0)} \left(\inf_{\tau \in \mathcal{S}_{0,T}(R)} P\big(V_\tau^{\tilde{v}_0, \pi} \geq U_\tau\big) \right). \qquad (7.2)$$

We prove that for any $\pi \in \mathcal{A}^{\mathbb{G}}(\tilde{v}_0)$ the worst-case scenario always occurs at $\tau \equiv T$, i.e.

$$\inf_{\tau \in \mathcal{S}_{0,T}(R)} P\big(V_\tau^{\tilde{v}_0, \pi} \geq U_\tau\big) = P\big(V_T^{\tilde{v}_0, \pi} \geq U_T\big). \qquad (7.3)$$

This result significantly simplifies (7.2), and without (7.3) we need to find a saddle point for the objective function $P\big(V_\tau^{\tilde{v}_0, \pi} \geq U_\tau\big)$. However, the existence of a saddle point is not always guaranteed, in particular for a stochastic dynamic problem such as (7.2). In a Black–Scholes framework, we solve (7.2) for its optimal value with equality.

We also show that the optimal trading strategy $\tilde{\pi}$ belongs to $\mathcal{A}^{\mathcal{F}}(\tilde{v}_0)$, and this helps us to give an explicit representation for the maximal probability of success and its optimal trading strategy. We study the reverse problem to the quantile hedging problem (7.2). This can be interpreted as finding the minimal initial cost $\tilde{v}_0 > 0$ for a given probability of a successful hedge $\alpha \in (0, 1)$.

We studied the guaranteed minimum maturity benefit (GMMB) equity-linked life insurance contracts. In this case, the policyholder is guaranteed the maximum of a predetermined amount (the guarantee) and an underlying stock index at the maturity time. If the guarantee matures in-the-money then the insurer is liable for the shortfall, otherwise the policyholder receives the stock index and the insurer's liability is zero.

The equity participation of the contract exposes the insurer to the market risk in terms of a European call/put option. By traditional actuarial approach, the mortality risk of the contract is usually managed by the client's survival probability over the term of the contract. In contrast to this approach, one can used the concept of random times and the enlargement of the filtrations to determine a hedging strategy that takes into account both financial risk and mortality risk dynamically.

In this chapter, we investigate another type of investment guarantee that resembles an American option. Using this connection and enlargement of filtration techniques, from the insurer's point of view, we will study the problem of maximizing the probability of a successful hedge under a capital constraint.

Definition 7.1. The guaranteed minimum death benefit (GMDB) is a type of equity-linked life insurance contract with two main characteristic features, investment opportunity and protection guarantee. Upon the insured's death during the term of the contract; if the underlying asset price rises then the insured enjoys the benefits of the equity investment, and in the case of a downside risk investment, the insurer guarantees a minimum payment to protect the insured against the market risk.

We consider contracts designed with a separate account format. This means that the insurer manages the fund available (from the premium) in the account by investing in the underlying equity, but the actual owner of the account is still the insured. Let F_t be the market value of the separate account and S_t be the price of the underlying equity investment at time t. Then

$$F_t = F_0 \frac{S_t}{S_0}(1-m)^t, \tag{7.4}$$

for $t = 0, 1, 2, ..., T$, where T is the maturity time and m denotes the management charge rate deducted from the account at the end of each month.

If the insured dies in the time interval $(t-1, t]$, for $t = 1, 2, ..., T$, the policyholder receives $\max(K, F_t)$. Mathematically speaking, a GMDB contract with guarantee $K > 0$ and the maturity time T has the following payoff:

$$\sum_{t=1}^{T} \max(K, F_t) \mathbb{1}_{\{t-1 < T(x) \leq t\}}, \tag{7.5}$$

where $T(x)$ represents the lifetime of a client who is currently at age x.

We suppose that the dynamic of the underlying asset price $(S_t)_{0 \leq t \leq T}$ is governed by the following Black–Scholes model:

$$\begin{cases} dS_t = S_t(\mu_t dt + \sigma_t dW_t); & S_0 > 0 \\ dB_t = B_t r_t dt; & B_0 = 1 \end{cases} \tag{7.6}$$

for $t \in [0, T]$. Here, $(W_t)_{0 \leq t \leq T}$ is a standard Brownian motion on the complete probability space $(\Omega, \mathcal{F} = (\mathcal{F}_t)_{0 \leq t \leq T} \subseteq \mathcal{G}, P)$, $(\mu_t)_{0 \leq t \leq T}$ and the positive process $(\sigma_t)_{0 \leq t \leq T}$ are $\mathcal{F}$-adapted processes representing the appreciation rate and the volatility of S, respectively. The nonnegative deterministic process $(r_t)_{0 \leq t \leq T}$ denotes the risk-free interest rate.

On the probability space $(\Omega, \mathcal{F} = (\mathcal{F}_t)_{0 \leq t \leq T} \subseteq \mathcal{G}, P)$, the equivalent martingale measure P^* for (7.6) is defined as:

$$\frac{dP^*}{dP} = \rho_T^*$$

where

$$\rho_t^{ast} = \exp\left(-\int_0^t \frac{\mu_s - r_s}{\sigma_s} dW_s - \int_0^t \frac{1}{2}\left(\frac{\mu_s - r_s}{\sigma_s}\right)^2 ds\right) \quad (7.7)$$

for all $t \in [0, T]$.

To satisfy the no-arbitrage condition and make the above model complete, we impose the following integrability conditions:

(1) $\int_0^T \left(\frac{\mu_s - r_s}{\sigma_s}\right)^2 ds < +\infty,$ (P-a.s.)

(2) $E[\rho_T^*] = 1$

Assumption 7.1. *For the sake of simplicity, we assume that $r_t \equiv 0$ with $\mu_t \equiv \mu \in \mathbb{R}$ and $\sigma_t \equiv \sigma > 0$ constant. In addition, we take the management rate $m \equiv 0$ and $F_0 = S_0$ in (7.4).*

In this setting, the GMDB contract (7.5) is simplified to:

$$H(D) := \sum_{t=1}^T \max(K, S_t) \mathbb{1}_{\{t-1 < T(x) \leq t\}} \quad (7.8)$$

In general, the mortality risk and the financial risk are not correlated. Hence it is natural to assume that:

Assumption 7.2. *$T(x)$ is a $\mathcal{G}$-measurable random variable independent of the risky asset S.*

Minimum guarantee equity-linked life insurance contracts are priced and hedged by a combination of actuarial methods and modern techniques of mathematical finance. By the law of large numbers, the mortality risk of the client is replaced by its expected value; and the financial risk associated to the underlying equity is managed by the methods of Black and Scholes.

As we already know, the Brennan and Schwartz hedging approach begins with the fact that minimum guarantee equity-linked life insurance contracts are separate account products. On the one hand, for each $t = 1, 2, \ldots, T$, we have $\max(K, S_t) = S_t + (K - S_t)^+$; and, on the other hand, the asset value S_t is available from the separate account. Therefore, the unhedged liability of the insurer for H(D), defined as in (7.8), can be written as follows:

$$\sum_{t=1}^T (K - S_t)^+ \mathbb{1}_{\{t-1 < T(x) \leq t\}}. \quad (7.9)$$

This means, using the Brennan and Schwartz approach, instead of (7.8) the following modified version of $H(D)$ is analyzed:

$$\hat{H} := \sum_{t=1}^{T} \mathrm{P}(t-1 < T(x) \leqslant t)(K - S_t)^+. \tag{7.10}$$

Let P* be the unique martingale probability measure of the Black–Scholes model (7.6). Then initial price of $\hat{H}$ is given by

$$\hat{H}(0) := \sum_{t=1}^{T} \mathrm{P}(t-1 < T(x) \leqslant t)\mathrm{E}^*\bigl[(K - S_t)^+\bigr] \tag{7.11}$$

where E^* denotes the expectation with respect to the probability measure P*.

If we want to adopt the above method, we need to hedge a weighted sum of multiple embedded put options $\sum_{t=1}^{T} \mathrm{P}(t-1 < T(x) \leqslant t)(K - S_t)^+$. This makes problem (7.2) tedious and complicated, even in a complete market setting. In contrast to the Brennan and Schwastz approach, we formulate problem (7.2) for $\sum_{t=1}^{t} \max(K, S_t) \mathbb{1}_{t-1 < T(x) \leqslant t}$ rather than $\sum_{t=1}^{T} \mathrm{P}(t-1 < T(x) \leqslant t)(K - S_t)^+$. The maximum property in $\max(K, S_t)$ is crucial to obtain the superhedging value process for $H(D)$ and to simplify the max-min problem (7.2) to a straightforward maximization problem. We only acknowledge the separate account format of the GMDB contracts after simplifying the quantile hedging problem (7.2).

As an alternative to the law of large numbers for dealing with the mortality risk, we construct a new filtration which enables us to consider $H(D)$ as an American option with only finitely many permitted exercise dates $\{1, 2, 3, \ldots, T\}$ in an incomplete market framework.

To do so, by progressively enlargement of filtrations, let us define

$$\mathcal{G}_t := \mathcal{F}_t \vee \mathcal{H}_t, \text{ for all } t \in [0, T],$$

where $\mathcal{H}_t := \sigma\bigl(T(x) \leqslant t\bigr)$ is the σ-field generated by $T(x)$ up to time $t \in [0, T]$. We denote the enlarged filtration $(\mathcal{G}_t)_{t \in [0,T]}$ by $\mathcal{G} := \mathcal{F} \vee \mathcal{H}$.

The financial model (7.6) equipped with the probability space $(\Omega, \mathcal{G}, \mathrm{P})$ and the new filtration $\mathcal{G} = (\mathcal{G}_t)_{t \in [0,T]}$ is an incomplete market because, for instance, $\mathbb{1}_{\{T(x) > T\}}$ is not attainable in this model. Assumption 7.2 implies that any $(\mathcal{F}, \mathrm{P})$-martingale remains a $(\mathcal{G}, \mathrm{P})$-martingale. This guarantees the no arbitrage condition in the new model.

Let us define a nonnegative function $\hat{\mu}$, known as the force of mortality or the hazard rate function, such that for all $t \geqslant 0$

$$_t p_x := P\bigl(T(x) > t\bigr) = \exp\left(-\int_0^t \hat{\mu}(x+s)ds\right).$$

We do not touch a mortality modeling, while an appropriate stochastic mortality modeling can bring reasonable advantages in such pricing methods.

As in Chapter 6 we introduce the set

$$\mathcal{D} := \left\{ (\kappa_t)_{0 \leqslant t \leqslant T} : \text{bounded}, \mathcal{G}\text{-predictable and } \kappa_t > -1 \ dt \times d\mathrm{P} \text{ a.e.} \right\}.$$

For any $\kappa \in \mathcal{D}$, define

$$\rho_t^\kappa = 1 + \int_0^t \kappa_s \rho_{s-}^\kappa dM_s, \quad t \in [0, T],$$

with $M_t = \mathbb{1}_{\{T(x) \leqslant t\}} - \int_0^{T(x) \wedge t} \hat{\mu}(x+s) ds$ which is a $(\mathcal{G}, \mathrm{P})$-martingale.

Using the definition of stochastic exponential, the unique solution to the above SDE is given by

$$\rho_t^\kappa = \left(1 + \kappa_{T(x)} \mathbb{1}_{\{T(x) \leqslant t\}}\right) \exp\left(-\int_0^{T(x) \wedge t} \kappa_s \hat{\mu}(x+s) ds\right).$$

Keeping in mind the above notations, we can provide an explicit representation for the Radon–Nikodym density of the martingale probability measures of $(S_t)_{t \in [0,T]}$ on $(\Omega, \mathcal{G} = (\mathcal{G}_t)_{0 \leqslant t \leqslant T} \subseteq \mathcal{G}, \mathrm{P})$ as follows:

$$\mathcal{Q} := \left\{ Q^\kappa \, \Big| \, \frac{dQ^\kappa}{d\mathrm{P}} = \rho_T^* \rho_T^\kappa \text{ for some } \kappa \in \mathcal{D} \right\}$$

To see the GMDB life insurance $H(D)$ defined in (7.8) as a Bermudan option, we give the next definition:

Definition 7.2. A Bermudan option is a particular type of American option which can be exercised only at a predetermined region of permitted exercise dates $R \subseteq [0,T]$. The payoff process is a nonnegative RCLL-adapted process denoted by $U = (U_t)_{t \in [0,T]}$ such that $U_t = 0$ for $t \notin R$. The option is exercised by choosing a $\mathcal{G}$-stopping time τ with values in R.

We consider $R = \{1, 2, 3, \ldots, T\}$, a suitable finite subset of $[0, T]$. The underlying Bermudan option with the payoff process

$$U = (U_t)_{t \in [0,T]} = \big(\max(K, S_t)\big)_{t \in [0,T]}$$

and the region of permitted exercise dates R is represented by the pair (U, R).

Remark 7.1. The exercise date of the GMDB contract $H(D)$ is not exactly $T(x)$. In fact, it is the smallest integer greater than or equal to $T(x)$. Let us denote this positive discrete time $\mathcal{G}_T$-random variable by $\tilde{\tau}$, we can use the ceiling function to represent $\tilde{\tau}$, i.e.,

$$\tilde{\tau} :=]T(x)[\tag{7.12}$$

Equity-linked life insurance contracts and Bermudan options   169

To view $H(D)$ as a Bermudan option, first we need to prove that the random exercise time $\tilde{\tau}$ is a $\mathcal{G}$-stopping time.

Lemma 7.1. *Let $\tilde{\tau}$ be defined as in (7.12), then $\tilde{\tau}$ is a $\mathcal{G}$-stopping time.*

Proof. For any $t \in [0, T]$, we have

$$\{\tilde{\tau} \leqslant t\} = \{\tilde{\tau} \leqslant [t]\} = \{T(x) \leqslant [t]\} \in \mathcal{G}_{[t]} \subseteq \mathcal{G}_t,$$

where $[t]$ is the floor function of t, the largest integer less than or equal to t. □

We consider the superhedging approach to price and hedge the Bermudan option

$$(U, R) = \Big(\big(\max(K, S_t) \big)_{t \in [0,T]}, R \Big).$$

By using optional decomposition of supermartingales, one can investigate this approach for the general case of American options. Moreover, one can utilize a backward argument to find an analytic formula for the superhedging value process of a Bermudan option. This technique is adapted here to price $H(D)$ on the probability space $\big(\Omega, \mathcal{G} = (\mathcal{G}_t)_{0 \leqslant t \leqslant T} \subseteq \mathcal{G}, P\big)$.

The superhedging value process of (U, R) at time $t \in [0, T]$ is defined as follows:

$$X_t := \operatorname*{ess\,sup}_{\substack{\kappa \in \mathcal{D} \\ \tau \in \mathcal{S}_{t,T}(R)}} \mathrm{E}^{Q^\kappa}\big[U_\tau | \mathcal{G}_t\big], \tag{7.13}$$

where $\mathcal{S}_{t,T}(R)$ is the set of all $\mathbb{G}$-stopping times with values in $R \cap [t, T]$, and $\mathrm{E}^{Q^\kappa}[\cdot]$ denotes expectation with respect to martingale probability measure Q^κ.

This method of backward arguments was proposed by Schweizer and it shows that X_{t_i} at $t_i \in R$ is equal to the $\mathcal{Q}$-uniform snell envelope of all the finite possible payoffs U_{t_j} for $j = i, i+1, ..., n$. In other words, for all $t_i \in R$, he drops the stopping times $\tau \in \mathcal{S}_{t_i,T}(R)$ in the calculation of X_{t_i}. Then X_t between two possible exercise dates t_i and t_{i+1} is determined by the price of a European option initiated at time t_i, maturity time t_{i+1} and the payoff $X_{t_{i+1}}$ at time t_{i+1}.

For the reader's convenience, we summarize Schweizer's method below.

Assume (U, R) to be a Bermudan option with $R = \{t_1, t_2, \ldots, t_n\} \subseteq [0, T]$ for some $n \in \mathbb{N}$ and $0 =: t_0 < t_1 < t_2 < \ldots < t_n = T$ such that

$$X_0 = \sup_{\substack{\kappa \in \mathcal{D} \\ i=1,2,\ldots,n}} \mathrm{E}^{Q^\kappa}\big[U_{t_i}\big] < +\infty.$$

Using a backward induction argument, let us define process $(B_i)_{i=0,1,2,\ldots,n}$ as follows

$$B_n := U_{t_n}$$

and
$$B_i := \max\left(U_{t_i}, \operatorname*{ess\,sup}_{\kappa \in \mathcal{D}} \mathrm{E}^{Q^\kappa}\left[B_{t_{i+1}}|\mathcal{G}_{t_i}\right]\right) \quad (7.14)$$

for $i = 0, 1, 2, ..., n-1$.

Then for the superhedging value process $(X_t)_{t\in[0,T]}$ introduced in (7.13) we have
$$X_{t_i} = B_i \ P\text{-a.s. for all } i = 0, 1, 2, \ldots, n.$$

Moreover, process $(X_t)_{t\in[0,T]}$ has an RCLL version on each subinterval (t_i, t_{i+1}) such that
$$X_t = X_{t_i+} + \int_{t_i}^t \pi_s^{(i)} dS_s - C_t^i \text{ for } t \in (t_i, t_{i+1}],$$

for some S-integrable $\mathbb{R}$-valued $\mathcal{G}$-predictable process $(\pi_t^{(i)})_{t\in(t_i, t_{i+1}]}$ and a nonnegative increasing $\mathcal{G}$-optional process $(C_t^{(i)})_{t\in(t_i, t_{i+1}]}$ with $C_{t_i}^{(i)} \equiv 0$. We set $X_{0+} := X_0$.

By attaching all the above n trading strategies $\pi^{(i)}$ and the consumption processes C^i, for all $t \in [0, T]$ we now define:
$$\pi_t := \sum_{i=0}^{n-1} \pi_t^{(i)} \mathbb{1}_{]t_i, t_{i+1}]} \quad (7.15)$$

and
$$C_t := \sum_{\substack{t_i \leq t \\ i=0,1,\ldots,n-1}} C_{t_i}^i + \sum_{i=0}^{n-1} C_t^i \mathbb{1}_{]t_i, t_{i+1}]} + \sum_{\substack{t_i < t \\ i=0,1,\ldots,n-1}} (X_{t_i} - X_{t_i+}). \quad (7.16)$$

Combining (7.15) and (7.16), we obtain
$$X_t = X_0 + \int_0^t \pi_s dS_s - C_t \text{ for } t \in [0, T]. \quad (7.17)$$

7.2 Quantile hedging of GMDB life insurance contracts

Let $v_0 > 0$ be a given initial capital, and $(\pi_t)_{t\in[0,T]}$ a $\mathcal{G}$-predictable S-integrable process. Then the self-financing value process $(V_t^{v_0,\pi})_{t\in[0,T]}$ corresponding to (v_0, π) is defined as follows:
$$V_t^{v_0,\pi} := v_0 + \int_0^t \pi_s dS_s, \ P\text{-a.s., for all } t \in [0, T].$$

If $V_t^{v_0,\pi} \geq 0$ P-a.s. for any $t \in [0,T]$, then the self-financing strategy (v_0, π) is called $\mathbb{G}$-admissible. The set of all $\mathcal{G}$-admissible trading strategies with the initial capital v_0 is denoted by $\mathcal{A}^{\mathcal{G}}(v_0)$.

We assume $\tilde{v}_0 > 0$, the available initial capital to superhedge the Bermudan option (U, R), is subject to the following constraint:

$$S_0 < \tilde{v}_0 < \sup_{\substack{\kappa \in \mathcal{D} \\ \tau \in \mathcal{S}_{0,T}(R)}} \mathrm{E}^{Q^\kappa}[U_\tau]. \qquad (7.18)$$

Remark 7.2. To buy the equity S at time $t = 0$ and hold it in a separate account, the insurer needs to have $\tilde{v}_0 \geq S_0$. However, if $\tilde{v}_0 = S_0$ then there will not be any initial capital available for the insurer to hedge its liability in the case of a guarantee ending up in-the-money. Therefore, in (7.18), it is natural to assume $S_0 < \tilde{v}_0$.

Since $\tilde{v}_0$ is strictly less than the initial cost of superhedging (U, R), for any choice of $\pi \in \mathcal{A}^{\mathcal{G}}(\tilde{v}_0)$, always there is the possibility of shortfall, i.e.,

$$\forall \pi \in \mathcal{A}^{\mathbb{G}}(\tilde{v}_0) \; \exists j \in \{1, ..., n\} \text{ s.t., } \mathrm{P}\left(\max(K, S_{t_j}) > V_{t_j}^{\tilde{v}_0, \pi}\right) > 0. \qquad (7.19)$$

We are looking for an optimal trading strategy $\tilde{\pi} \in \mathcal{A}^{\mathcal{G}}(\tilde{v}_0)$ such that it minimizes the worst possible scenario of a shortfall risk as described in (7.19). Equivalently, we formulate the quantile hedging problem for the Bermudan option (U, R) with the available initial capital $\tilde{v}_0$ as follows:

$$\sup_{\pi \in \mathcal{A}^{\mathcal{G}}(\tilde{v}_0)} \left(\inf_{\tau \in \mathcal{S}_{0,T}(R)} \mathrm{P}(V_\tau^{\tilde{v}_0, \pi} \geq U_\tau) \right) \qquad (7.20)$$

This max-min problem is a stochastic game between the time of death of the insured and the insurer's trading strategy to hedge (U, R). From the insurer's point of view, by problem (7.20) we want to maximize the worst probability of a successful hedge over all permitted exercise dates in R.

As the first step to deal with problem (7.20), we exploit the Schweizer's method to determine the superhedging value process X_t introduced in (7.20). The next lemma gives us a tool to compute the underlying $\mathcal{G}_t$-conditional expectations in the definition of B_is, in (7.14), in terms of $\mathcal{F}_t$-conditional expectations.

Lemma 7.2. *Let H be an $\mathcal{F}_T$-measurable random variable. Then, for any $\kappa \in \mathcal{D}$ and $t \in [0,T]$, we have*

$$\mathrm{E}^{Q^\kappa}[H|\mathcal{G}_t] = \mathrm{E}^*[H|\mathcal{F}_t]. \qquad (7.21)$$

Proof. Using Bayes formula, one can write

$$\mathrm{E}^{Q^\kappa}[H|\mathcal{G}_t] = \frac{1}{\rho_t^\kappa \rho_t^*} \mathrm{E}[\rho_T^\kappa \rho_T^* H | \mathcal{F}_t], \text{ for any } \kappa \in \mathcal{D}.$$

$(\mathrm{E}[\rho_T^* H | \mathcal{F}_u])_{0 \leq u \leq T}$ is an $(\mathcal{F}, \mathrm{P})$-martingale which, due to Assumption 7.2, follows a $(\mathcal{G}, \mathrm{P})$-martingale too. In addition, $(\rho_u^\kappa)_{0 \leq u \leq T}$ is a $(\mathcal{G}, \mathrm{P})$-martingale orthogonal to $(\mathrm{E}[\rho_T^* H | \mathcal{F}_u])_{0 \leq u \leq T}$, since their quadratic covariation is equal to zero. This implies their product $(\rho_u^\kappa \mathrm{E}[\rho_T^* H | \mathcal{F}_u])_{0 \leq u \leq T}$ is a $(\mathcal{G}, \mathrm{P})$-local martingale. By passing through $(H \wedge m)_{m \geq 1}$ and using the monotone convergence theorem, we get

$$\mathrm{E}\big[\rho_T^\kappa \mathrm{E}[\rho_T^* H | \mathcal{F}_T] \big| \mathcal{G}_t\big] = \rho_t^\kappa \mathrm{E}[\rho_T^* H | \mathcal{F}_t].$$

Hence,

$$\frac{1}{\rho_t^\kappa \rho_t^*} \mathrm{E}[\rho_T^\kappa \rho_T^* H | \mathcal{G}_t] = \frac{1}{\rho_t^*} \mathrm{E}[\rho_T^* H | \mathcal{F}_t],$$

from this equation, we can easily derive (7.21). □

In the following lemma we compute the process $(X_t)_{t \in [0,T]}$:

Lemma 7.3. *Let $(X_t)_{t \in [0,T]}$ be the superhedging value process of the Bermudan option $(U, R) = \big((\max(K, S_t))_{t \in [0,T]}, R\big)$. Then we have*

$$X_t = S_t + \mathrm{E}^*\big[(K - S_T)^+ \big| \mathcal{F}_t\big], \quad \text{for all } t \in [0, T].$$

In particular, the initial cost of superhedging is given by

$$X_0 = S_0 + \mathrm{E}^*\big[(K - S_T)^+\big].$$

Proof. By the definition of B_n:

$$B_n = U_{t_n} = \max(K, S_{t_n}).$$

From this and (7.14), we obtain

$$\begin{aligned} B_{n-1} &= \max\Big(U_{t_{n-1}}, \operatorname*{ess\,sup}_{\kappa \in \mathcal{D}} \mathrm{E}^{Q^\kappa}[B_n | \mathcal{G}_{t_{n-1}}]\Big) \\ &= \max\Big(\max(K, S_{t_{n-1}}), \mathrm{E}^*\big[\max(K, S_{t_n}) \big| \mathcal{F}_{t_{n-1}}\big]\Big), \end{aligned} \quad (7.22)$$

where to get the second equality we have used Lemma 7.2. On the other hand, it is easy to see that

$$\mathrm{E}^*\big[\max(K, S_{t_n}) \big| \mathcal{F}_{t_{n-1}}\big] \geq K$$

and

$$\mathrm{E}^*\big[\max(K, S_{t_n}) \big| \mathcal{F}_{t_{n-1}}\big] \geq \mathrm{E}^*\big[S_{t_n} \big| \mathcal{F}_{t_{n-1}}\big] = S_{t_{n-1}}.$$

Hence, (7.22) becomes

$$\begin{aligned} B_{n-1} &= \mathrm{E}^*\big[\max(K, S_{t_n}) \big| \mathcal{F}_{t_{n-1}}\big] = \mathrm{E}^*\big[S_{t_n} + (K - S_{t_n})^+ \big| \mathcal{F}_{t_{n-1}}\big] \\ &= S_{t_{n-1}} + \mathrm{E}^*\big[(K - S_{t_n})^+ \big| \mathcal{F}_{t_{n-1}}\big]. \end{aligned}$$

By induction, we can show that for all $i = 0, 1, \ldots, n-1$

$$B_i = \max\left(U_{t_i}, \operatorname*{ess\,sup}_{\kappa \in \mathcal{D}} \mathrm{E}^{q^\kappa}\left[B_{i+1}|\mathcal{G}_{t_i}\right]\right)$$

$$= \max\left(\max(K, S_{t_i}), \operatorname*{ess\,sup}_{\kappa \in \mathcal{D}} \mathrm{E}^{Q^\kappa}\left[\mathrm{E}^*\left[\max(K, S_{t_n})|\mathcal{F}_{t_{i+1}}\right]|\mathcal{G}_{t_i}\right]\right) \quad (7.23)$$

$$= \max\left(\max(K, S_{t_i}), \mathrm{E}^*\left[\max(K, S_{t_n})|\mathcal{F}_{t_i}\right]\right)$$

$$= \mathrm{E}^*\left[S_{t_n} + (K - S_{t_n})^+|\mathcal{F}_{t_i}\right] = S_{t_i} + \mathrm{E}^*\left[(K - S_{t_n})^+|\mathcal{F}_{t_i}\right]$$

In this case, we applied Lemma 7.2 on $H = \mathrm{E}^*\left[\max(K, S_{t_n})|\mathcal{F}_{t_{i+1}}\right]$ with the fact that $\mathcal{F}_{t_i} \subseteq \mathcal{F}_{t_{i+1}}$. In particular, as a side product of (7.23), we can see that $(B_i)_{i=0,1,\ldots,n}$ is an $(\mathcal{F}, \mathrm{P}^*)$-martingale.

Having B_is determined, by Lemma 7.2 we now calculate X_t for $t \in (t_i, t_{i+1}]$ for each $i = 0, 1, 2, \ldots, n-1$ as follows:

$$X_t = \operatorname*{ess\,sup}_{\kappa \in \mathcal{D}} \mathrm{E}^{Q^\kappa}\left[S_{t_{i+1}} + \mathrm{E}^*\left[(K - S_{t_n})^+|\mathcal{F}_{t_{i+1}}\right]|\mathcal{G}_t\right]$$

$$= \mathrm{E}^*\left[S_{t_{i+1}} + \mathrm{E}^*\left[(K - S_{t_n})^+|\mathcal{F}_{t_{i+1}}\right]|\mathcal{F}_t\right] = S_t + \mathrm{E}^*\left[(K - S_{t_n})^+|\mathcal{F}_t\right]$$

Since $\left(S_t + \mathrm{E}^*\left[(K - S_{t_n})^+|\mathcal{F}_t\right]\right)_{t \in [0,T]}$ is a continuous process and consumption process $C_t^i \equiv 0$ on each subinterval $(t_i, t_{i+1}]$, it is clear that the overall consumption process $(C_t)_{t \in [0,T]}$ defined by (7.16) is zero.

Combining the above with (7.15)–(7.17) completes the proof. $\square$

Now we come back to the quantile hedging problem (7.20). We solve problem (7.20) for its precise optimal value. In addition, explicit form solutions will be provided for the maximal probability and the optimal hedge achieving this value.

Theorem 7.1. *Consider the quantile hedging problem (7.20). We have*

$$\sup_{\pi \in \mathcal{A}^{\mathcal{G}}(\tilde{v}_0)} \left(\inf_{\tau \in \mathcal{S}_{0,T}(R)} \mathrm{P}(V_\tau^{\tilde{v}_0, \pi} \geq U_\tau)\right) = \sup_{\pi \in \mathcal{A}^{\mathcal{G}}(\tilde{v}_0)} \mathrm{P}(V_T^{\tilde{v}_0, \pi} \geq U_T)$$

Proof. For an arbitrary $\pi \in \mathcal{A}^{\mathcal{G}}(\tilde{v}_0)$, let $A_T := \{V_T^{\tilde{v}_0, \pi} \geq U_T\} \in \mathcal{G}_T$. Then

$$\mathrm{E}^{Q^\kappa}\left[V_T^{\tilde{v}_0, \pi} \mathbb{1}_{A_T}|\mathcal{G}_{t_i}\right] \geq \mathrm{E}^{Q^\kappa}\left[U_T \mathbb{1}_{A_T}|\mathcal{G}_{t_i}\right], \quad (7.24)$$

for all $\kappa \in \mathcal{D}$ and $i = 0, 1, 2, \ldots, n$. Since $\left(V_t^{\tilde{v}_0, \pi}\right)_{t \in [0,T]}$ is a $(\mathbb{G}, Q^\kappa)$-supermartingale, we get

$$V_{t_i}^{\tilde{v}_0, \pi} \geq \mathrm{E}^{Q^\kappa}\left[V_T^{\tilde{v}_0, \pi}|\mathcal{G}_{t_i}\right] \geq \mathrm{E}^{Q^\kappa}\left[V_T^{\tilde{v}_0, \pi} \mathbb{1}_{A_T}|\mathcal{G}_{t_i}\right]. \quad (7.25)$$

On the other hand, Lemma 7.2 implies

$$\mathrm{E}^{Q^\kappa}\big[U_T \mathbb{1}_{A_T}\big|\mathcal{G}_{t_i}\big] + \mathrm{E}^{Q^\kappa}\big[U_T \mathbb{1}_{A_T^c}\big|\mathcal{G}_{t_i}\big] = \mathrm{E}^{Q^\kappa}\big[U_T\big|\mathcal{G}_{t_i}\big]$$
$$= \mathrm{E}^*\big[U_T\big|\mathcal{F}_{t_i}\big] = \mathrm{E}^*\big[B_n\big|\mathcal{F}_{t_i}\big] = B_i \geqslant U_{t_i}, \quad (7.26)$$

where we used the martingale property of $(B_i)_{i=0,1,\ldots,n}$ from the proof of Lemma 7.3, and also the fact that by the definition B_i dominates U_i. Combining (7.24)–(7.26), we can see that

$$V_{t_i}^{\tilde{v}_0,\pi} \geqslant U_{t_i} - \mathrm{E}^{Q^\kappa}\big[U_T \mathbb{1}_{A_T^c}\big|\mathcal{G}_{t_i}\big]. \quad (7.27)$$

Lemma 7.4. *Let $s \leqslant t$ and $A \in \mathcal{G}_t$. Then for any nonnegative $\mathcal{G}_t$-measurable random variable Y, we have*

$$\mathrm{E}\big[Y \mathbb{1}_{A^c}\big|\mathcal{G}_s\big]\mathbb{1}_A = 0, \quad \text{P-a.s.} \quad (7.28)$$

Proof. We prove this lemma by contradiction. Let

$$B := \big\{\omega \in A \,:\, \mathrm{E}\big[Y\mathbb{1}_{A^c}\big|\mathcal{G}_s\big] \neq 0\big\}.$$

Then, by $B \in \mathcal{G}_s$, we get

$$0 < \int_B \mathrm{E}\big[Y\mathbb{1}_{A^c}\big|\mathcal{G}_s\big]dP = \int_B Y\mathbb{1}_{A^c}dP = \int_{B \cap A^c} Y dP.$$

On the other hand, $B \subseteq A$ implies $\int_{B \cap A^c} Y dP = 0$. Therefore B must be P almost surely empty and (7.28) is satisfied. $\square$

Multiplying both sides of the inequality (7.27) by $\mathbb{1}_{A_T}$ and then applying Lemma 7.4 to $\mathrm{E}^{Q^\kappa}\big[U_T \mathbb{1}_{A_T^c}\big|\mathcal{G}_{t_i}\big]\mathbb{1}_{A_T}$, we have shown that

$$V_{t_i}^{\tilde{v}_0,\pi} \mathbb{1}_{A_T} \geqslant U_{t_i} \mathbb{1}_{A_T}.$$

Therefore, on $A_T = \{V_T^{\tilde{v}_0,\pi} \geqslant U_T\} \in \mathcal{G}_T$, one can write

$$V_{t_i}^{\tilde{v}_0,\pi} \mathbb{1}_{\{\tau=t_i\}} \geqslant U_{t_i} \mathbb{1}_{\{\tau=t_i\}}, \quad \text{for any} \quad \tau \in \mathcal{S}_{0,T}(R).$$

This leads us to

$$V_\tau^{\tilde{v}_0,\pi} = \sum_{i=0}^n V_{t_i}^{\tilde{v}_0,\pi} \mathbb{1}_{\{\tau=t_i\}} \geqslant \sum_{i=0}^n U_{t_i} \mathbb{1}_{\{\tau=t_i\}} = U_\tau.$$

Hence we obtain

$$\{V_T^{\tilde{v}_0,\pi} \geqslant U_T\} \subseteq \{V_\tau^{\tilde{v}_0,\pi} \geqslant U_\tau\}, \quad \text{for any} \quad \tau \in \mathcal{S}_{0,T}(R),$$

and consequently we can write:

$$\inf_{\tau \in \mathcal{S}_{0,T}(R)} \mathrm{P}\big(V_\tau^{\tilde{v}_0,\pi} \geqslant U_\tau\big) = \mathrm{P}\big(V_T^{\tilde{v}_0,\pi} \geqslant U_T\big).$$

This completes the proof of Theorem 7.1. $\square$

By Theorem 7.1, the quantile hedging problem of Bermudan option (U, R) is reduced to its corresponding problem for the European option $U_T = \max(K, S_T)$. Let us recall that a GMDB contract is managed in a separate account format. This means the underlying asset $(S_t)_{t \in [0,T]}$ is held in a separate account, and, at time $t = T$, the unhedged liability of the insurer is the put option $(K - S_T)^+$ not $\max(K, S_T) = S_T + (K - S_T)^+$.

Keeping in mind the above discussion and the initial capital constraint (7.18), first we set aside S_0 to buy the equity at time $t = 0$ for the separate account. Then we consider $\hat{v}_0 := \tilde{v}_0 - S_0 > 0$ as the available initial capital to determine an optimal trading strategy for maximizing probability of success for $(K - S_T)^+$. Taking into account this argument and Theorem 7.1, we simplify problem (7.20) as follows:

$$\sup_{\pi \in \mathcal{A}^{\mathcal{G}}(\hat{v}_0)} P\left(V_T^{\hat{v}_0, \pi} \geq (K - S_T)^+\right). \tag{7.29}$$

Moreover, using (7.18) and Lemma 7.3, the new initial capital $\hat{v}_0$ is under the constraint:

$$0 < \hat{v}_0 < E^*\left[(K - S_T)^+\right].$$

Notice that the maximization problem (7.29) runs over trading strategies $\pi \in \mathcal{A}^{\mathcal{G}}(\hat{v}_0)$. We utilize the Neyman–Pearson lemma, Assumption 7.2 and a decomposition theorem for $\mathcal{G}$-adapted processes to replace $\mathcal{A}^{\mathcal{G}}(\hat{v}_0)$ with $\mathcal{A}^{\mathcal{F}}(\hat{v}_0)$.

Theorem 7.2. *Let Y be a nonnegative $\mathcal{F}_T$-measurable random variable and $\hat{v}_0 > 0$. Then we have*

$$\sup_{\pi \in \mathcal{A}^{\mathcal{G}}(\hat{v}_0)} P\left(V_T^{\hat{v}_0, \pi} \geq Y\right) = \sup_{\pi \in \mathcal{A}^{\mathcal{F}}(\hat{v}_0)} P\left(V_T^{\hat{v}_0, \pi} \geq Y\right).$$

Proof. Let $\hat{A} \in \mathcal{G}_T$ be a solution to the problem

$$\max_{A \in \mathcal{G}_T} P(A) \tag{7.30}$$

subject to the constraint

$$\sup_{\kappa \in \mathcal{D}} E^{Q^\kappa}[Y \mathbb{1}_A] \leq \hat{v}_0.$$

Then the trading strategy $\hat{\pi} \in \mathcal{A}^{\mathcal{G}}(\hat{v}_0)$ obtained from the optional decomposition of

$$\operatorname*{ess\,sup}_{\kappa \in \mathcal{D}} E^{Q^\kappa}\left[Y \mathbb{1}_{\hat{A}} \mid \mathcal{G}_t\right]$$

solves the problem

$$\sup_{\pi \in \mathcal{A}^{\mathcal{G}}(\hat{v}_0)} P\left(V_T^{\hat{v}_0, \pi} \geq Y\right).$$

Moreover, the maximal probability of success is given by:
$$P(V_T^{\hat{v}_0,\hat{\pi}} \geq Y) = P(\hat{A}). \tag{7.31}$$

Similarly, suppose $\tilde{A} \in \mathcal{F}_T$ is a solution to the following problem:
$$\max_{A \in \mathcal{F}_T} P(A)$$

subject to the constraint
$$E^*[Y\mathbb{1}_A] \leq \hat{v}_0.$$

Then $\tilde{\pi} \in \mathcal{A}^{\mathcal{F}}(\hat{v}_0)$ as the perfect hedge of the modified claim $Y\mathbb{1}_{\tilde{A}}$ solves the problem
$$\sup_{\pi \in \mathcal{A}^{\mathcal{F}}(\hat{v}_0)} P(V_T^{\hat{v}_0,\pi} \geq Y), \tag{7.32}$$

and the maximum probability is equal to:
$$P(V_T^{\hat{v}_0,\tilde{\pi}} \geq Y) = P(\tilde{A}). \tag{7.33}$$

Since $\mathcal{A}^{\mathcal{F}}(\hat{v}_0) \subseteq \mathcal{A}^{\mathcal{G}}(\hat{v}_0)$, it is easy to see that
$$P(\tilde{A}) \leq P(\hat{A}). \tag{7.34}$$

To finish the proof, we establish the reverse inequality. The $\mathcal{G}$-predictable process $(\hat{\pi}_t)_{t \in [0,T]} \in \mathcal{A}^{\mathcal{G}}(\hat{v}_0)$ admits the following decomposition:
$$\hat{\pi}_t = \hat{\pi}_t^{\mathcal{F}} \mathbb{1}_{\{t \leq T(x)\}} + \hat{\pi}_t^d(T(x))\mathbb{1}_{\{t > T(x)\}}, \quad \text{for } t \in [0,T], \tag{7.35}$$

where

(i) $(\hat{\pi}_t^{\mathcal{F}})_{t \in [0,T]}$ is an $\mathcal{F}$-predictable process.

(ii) $\{(\hat{\pi}_t^d(\theta))_{\theta \leq t \leq T} : \text{ for } \theta \in [0,T]\}$ is a family of $\mathcal{F}_t \times \mathcal{B}(\mathbb{R}_+)$-measurable functions.

(iii) For any fixed $\theta \in [0,T]$, process $(\hat{\pi}_t^d(\theta))_{\theta \leq t \leq T}$ is $\mathcal{F}$-predictable.

Decomposition (7.35) implies
$$V_T^{\hat{v}_0,\hat{\pi}} = V_T^{\hat{v}_0,\hat{\pi}^{\mathcal{F}}} \mathbb{1}_{\{T < T(x)\}} + V_T^{\hat{v}_0,\hat{\pi}^d(\theta)} \mathbb{1}_{\{T \geq \theta\}}\big|_{\theta = T(x)}.$$

By (7.32) and (7.33), we get
$$P(V_T^{\hat{v}_0,\hat{\pi}^{\mathcal{F}}} \geq Y) \leq P(\tilde{A}). \tag{7.36}$$

With a similar argument, for any $\theta \in [0,T]$
$$P(V_T^{\hat{v}_0,\hat{\pi}^{\mathcal{F}}(\theta)} \geq Y) \leq P(\tilde{A}).$$

As in Chapter 6 for computing expectation involving random times, we can write

$$P(V_T^{\hat{v}_0,\hat{\pi}^{\mathcal{F}}(\theta)} \geq Y)P(T(x) = \theta) \leq P(\tilde{A})P(T(x) = \theta)$$

$$\int_0^T P(V_T^{\hat{v}_0,\hat{\pi}^{\mathcal{F}}(\theta)} \geq Y)P(T(x) = \theta)d\theta \leq \int_0^T P(\tilde{A})P(T(x) = \theta)d\theta$$

$$E\left[\int_0^T \mathbb{1}_{\{V_T^{\hat{v}_0,\hat{\pi}^{\mathcal{F}}(\theta)} \geq Y\}} P(T(x) = \theta)d\theta\right] \leq P(\tilde{A}) \int_0^T P(T(x) = \theta)d\theta$$

$$P\left(\{V_T^{\hat{v}_0,\hat{\pi}^{\mathcal{F}}(T(x))} \geq Y\} \cap \{T(x) \leq T\}\right) \leq P(\tilde{A})P(T(x) \leq T) \quad (7.37)$$

By multiplying both sides of (7.36) by $P(T(x) > T)$ and then combining with (7.37) and (7.31), we have

$$P(\hat{A}) = P(V_T^{\hat{v}_0,\hat{\pi}} \geq Y) = P\left(\{V_T^{\hat{v}_0,\hat{\pi}^{\mathcal{F}}} \geq Y\} \cap \{T(x) > T\}\right)$$
$$+ P\left(\{V_T^{\hat{v}_0,\hat{\pi}^{\mathcal{F}}(T(x))} \geq Y\} \cap \{T(x) \leq T\}\right) \leq P(\tilde{A})$$

Therefore the reverse inequality of (7.33) is proved, and this means $\tilde{A} = \{V_T^{\hat{v}_0,\hat{\pi}} \geq Y\}$ solves problem (7.30). As an immediate consequence, $\tilde{\pi} \in \mathcal{A}^{\mathcal{F}}(\hat{v}_0)$ is a solution to problem (7.29) and the proof is finished. $\square$

In the next theorem, we provide an explicit form solution for the quantile hedging problem (7.29):

Theorem 7.3. *For a given initial capital $\hat{v}_0 > 0$, consider the optimization problem (7.29). Define European option $\tilde{H}$ as follows*

$$\tilde{H} := (K - S_T)^+ - (\tilde{K} - S_T)^+ - (K - \tilde{K})\mathbb{1}_{\{S_T \leq K\}},$$

where $\tilde{K} \in (0, K)$ is a constant subject to the constraint

$$E^*[\tilde{H}] = \hat{v}_0.$$

Then perfect trading strategy $(\tilde{\pi}_t)_{t \in [0,T]} \in \mathcal{A}^{\mathcal{F}}(\hat{v}_0)$ for $\tilde{H}$ solves problem (7.29), and the maximal probability of success is given by:

$$P(V_T^{\hat{v}_0,\tilde{\pi}} \geq (K - S_T)^+) = \Phi\left(\frac{\ln \tilde{S}_0 - \ln K}{\sigma\sqrt{T}} - \frac{1}{2}\sigma\sqrt{T}\right),$$

where Φ is the standard normal distribution function.

Proof. Keeping in mind Theorem 7.2, define

$$\tilde{A} := \left\{ \frac{d\mathrm{P}}{d\mathrm{P}^*} > const.\,(K - S_T)^+ \right\}.$$

As we know from the theory of quantile hedging, the replicating strategy $(\tilde{\pi}_t)_{t\in[0,T]} \in \mathcal{A}^{\mathcal{F}}(\hat{v}_0)$ for the modified claim $(K - S_T)^+ \mathbb{1}_{\tilde{A}}$ is an optimal solution to the following problem

$$\sup_{\pi \in \mathcal{A}^{\mathcal{F}}(\hat{v}_0)} \mathrm{P}\big(V_T^{\hat{v}_0,\pi} \geqslant (K - S_T)^+\big)$$

with the maximal probability of success

$$\mathrm{P}\big(V_T^{\hat{v}_0,\tilde{\pi}} \geqslant (K - S_T)^+\big) = \mathrm{P}(\tilde{A}).$$

By the definition of P* from (7.7) and the unique solution to SDE (7.6), we can rewrite $\tilde{A}$ as follows

$$\tilde{A} = \left\{ S_T^{\frac{\mu}{\sigma^2}} > \tilde{\lambda}(K - S_T)^+ \right\}, \tag{7.38}$$

where $\tilde{\lambda}$ is a positive constant to be determined from

$$\mathrm{E}^*\big[(K - S_T)^+ \mathbb{1}_{\tilde{A}}\big] = \hat{v}_0. \tag{7.39}$$

Regardless of $\frac{\mu}{\sigma^2} \leqslant 1$ or $\frac{\mu}{\sigma^2} > 1$, the increasing function $s^{\frac{\mu}{\sigma^2}}$ intersects with $\tilde{\lambda}(K - s)^+$ at exactly one point $\tilde{K} \in (0, K)$. Hence, from (7.38), we obtain

$$\tilde{A} = \{S_T > \tilde{K}\} = \left\{ S_0 \exp\left(\sigma W_T + \left(\mu - \frac{1}{2}\sigma^2\right)T\right) > \tilde{K} \right\}$$
$$= \left\{ S_0 \exp\left(\sigma W_T^* - \frac{1}{2}\sigma^2 T\right) > \tilde{K} \right\} = \{W_T^* > \bar{K}\}, \tag{7.40}$$

where $\bar{K} > 0$ is a constant, and $(W_t^*)_{t\in[0,T]} := \big(W_t + \frac{\mu}{\sigma}t\big)_{t\in[0,T]}$ is a standard $(\mathcal{F}, \mathrm{P}^*)$-Brownian motion by Girsanov's theorem.

To exploit the Black–Scholes formula, we represent the modified claim $(K - S_T)^+ \mathbb{1}_{\tilde{A}}$ as follows

$$(K - S_T)^+ \mathbb{1}_{\tilde{A}} = (K - S_T)^+ - (\tilde{K} - S_T)^+ + (\tilde{K} - K)\mathbb{1}_{\{S_T \leqslant \tilde{K}\}}$$

. By this and (7.39), constant $\tilde{K}$ can be computed from

$$\hat{v}_0 = \mathrm{E}^*\big[(K - S_T)^+ \mathbb{1}_{\tilde{A}}\big] = K\Phi(-d_-(K)) - S_0\Phi(-d_+(K))$$
$$- \tilde{K}\Phi(-d_-(\tilde{K})) + S_0\Phi(-d_+(\tilde{K})) + (\tilde{K} - K)\Phi(-d_-(\tilde{K}))$$
$$= K\Big[\Phi(-d_-(K)) - \Phi(-d_-(\tilde{K}))\Big] - S_0\Big[\Phi(-d_+(K)) - \Phi(-d_+(\tilde{K}))\Big]$$

where for any $z > 0$, $d_\pm(z)$ is defined as follows

$$d_\pm(z) := \frac{\ln S_0 - \ln z}{\sigma\sqrt{T}} \pm \frac{1}{2}\sigma\sqrt{T}.$$

By having $\tilde{K}$ from (7.39) and (7.40), then the maximal probability of success is given by

$$P(\tilde{A}) = P\left(S_0 \exp\left(\sigma W_T + (\mu - \frac{1}{2}\sigma^2)T\right) > \tilde{K}\right) = \Phi(d_-(\tilde{K})).$$

□

Let us study the inverse problem to the max-min problem (7.2). To be precise, for a given probability of success $\alpha \in (0,1)$, we find the minimum initial capital $\tilde{v}_0$ such that for some $\pi \in \mathcal{A}^{\mathcal{G}}(\tilde{v}_0)$ the following inequality holds true:

$$\inf_{\tau \in \mathcal{S}_{0,T}(R)} P(V_T^{\tilde{v}_0,\pi} \geq U_\tau) \geq \alpha. \tag{7.41}$$

Using Theorems 7.1 and 7.2, we can rewrite (7.41) as below:

$$\begin{aligned}\tilde{v}_0 &= \inf\{v > 0 \mid P(V_T^{v,\pi} \geq U_T) \geq \alpha \text{ for some } \pi \in \mathcal{A}^{\mathcal{G}}(v)\} \\ &= \inf\{v > 0 \mid P(V_T^{v,\pi} \geq U_T) \geq \alpha \text{ for some } \pi \in \mathcal{A}^{\mathcal{F}}(v)\},\end{aligned} \tag{7.42}$$

where the second equality is obtained by Theorem 7.2 which allows us to replace $\mathcal{A}^{\mathcal{G}}(v)$ with $\mathcal{A}^{\mathcal{F}}(v)$. By a similar argument as in (7.30)–(7.34), problem (7.42) can be formulated as the following minimization problem:

$$\min_{A \in \mathcal{F}_T} E^*[U_T \mathbb{1}_A] \tag{7.43}$$

under the constraint

$$P(A) \geq \alpha. \tag{7.44}$$

Instead of solving problem (7.43)–(7.44), we solve its equivalent problem defined as below:

$$\max_{B \in \mathcal{F}_T} E^*[U_T \mathbb{1}_B] \tag{7.45}$$

under the constraint

$$P(B) \leq 1 - \alpha. \tag{7.46}$$

It is easy to see that if $\tilde{B} \in \mathcal{F}_T$ is a solution to problem (7.45)–(7.46) then $\tilde{A} := \tilde{B}^c$ solves (7.43)–(7.44). As we discussed before in Theorem 7.2, by the GMDB contract design, at time T the unhedged liability of the insurer is the embedded put option $(K - S_T)^+$ rather than $U_T = \max(K, S_T) = S_T + (K - S_T)^+$. Taking into account this discussion, we simplify problem (7.40)–(7.46) as follows:

$$\max_{B \in \mathcal{F}_T} E^*[(K - S_T)^+ \mathbb{1}_B] \tag{7.47}$$

under the constraint
$$P(B) \leq 1 - \alpha. \tag{7.48}$$

Henceforth, by the optimal hedge to minimize the initial cost, we mean the minimal initial cost associated to $(K - S_T)^+$. Since S follows a geometric Brownian motion model and P^* is given by (7.7), similar to the proof of Theorem 7.3, by the Neyman–Pearson lemma we can show that there exists $\tilde{B} \in \mathcal{A}_T$ which solves problem (7.47)–(7.48). The above discussion is summarized in Theorem 7.4.

Theorem 7.4. *For a given probability of success $\alpha \in (0,1)$, let $\tilde{A} \in \mathcal{F}_T$ be a solution to the following minimization problem*

$$\min_{A \in \mathcal{F}_T} E^*[(K - S_T)^+ \mathbb{1}_A]$$

under the constraint

$$P(A) \geq \alpha.$$

Then we have:

(1) $\tilde{A} = \{S_T > \hat{K}\}$ for a positive constant $\hat{K}$ to be determined from

$$\Phi\left(\frac{\ln S_0 - \ln \hat{K}}{\sigma\sqrt{T}} - \frac{1}{2}\sigma\sqrt{T}\right) = \alpha.$$

(2) The minimal initial cost associated to problem (7.41) is given by

$$\hat{v}_0 = E^*\left[(K - S_T)^+ \mathbb{1}_{\{S_T > \hat{K}\}}\right].$$

Moreover, the perfect hedge $(\tilde{\pi}_t)_{t \in [0,T]} \in \mathcal{A}^{\mathcal{F}}(\hat{v}_0)$ to replicate $(K - S_T)^+ \mathbb{1}_{\{S_T > \hat{K}\}}$ is the optimal hedge corresponding to the cost minimization problem (7.41).

Proof. By the Neyman–Pearson lemma, the optimal solution to problem (7.47)-(7.48) is given by

$$\tilde{B} = \left\{\frac{dP}{dP^*} < const. \, (K - S_T)^+\right\}.$$

Similar to (7.40), this can be simplified to

$$\tilde{B} = \left\{S_T < \hat{K}\right\},$$

where the positive constant $\hat{K}$ is computed from the constraint

$$P(S_T < \hat{K}) = 1 - \alpha.$$

By the arguments before Theorem 7.4 and the fact that $P(S_T = \hat{K}) = 0$, the first part of the theorem is proved. Part (2) of the theorem is derived analogous to Theorem 7.3. □

Remark 7.3. Considering a Black–Scholes model, we define a guaranteed minimum death benefit (GMDB) life insurance contract in this market. The main aim of this chapter is to solve the quantile hedging problem (7.20) under an initial capital constraint. To do so, we progressively enlarge the filtration generated by the underlying asset with the filtration generated by the survival process of the insured. Then the GMDB contract is considered a Bermudan option on the probability space equipped with the enlarged filtration. The independency assumption between the mortality risk and the financial risk combined with the minimum guarantee structure of the payment simplifies the superhedging method into a perfect hedge in the original complete market. Moreover, the max-min problem corresponding to the quantile hedging problem of the GMDB contract in the enlarged filtration is converted into a straightforward quantile hedging problem for a European put option in the filtration generated by the Black–Scholes model.

7.3 Numerical illustrations

In the following, we provide numerical examples for the results obtained in Section 7.2. and compare our method, the Bermudan option hedging method, with the traditional hedging method described in Section 7.1. Throughout this section, let $x = 50$, the insured is 50 years old, $\mu = 0.04$, $\sigma = 0.2$ per year (equivalently $\mu = 0.0033$, $\sigma = 0.0577$ per month), $S_0 = 100$, and $K = 80$.

Maturity T (years)	Total Initial Capital	Bermudan Option Method	Traditional Hedging Method
5	$2.34	72.61%	53.88%
10	$5.98	75.07%	46.97%
15	$9.33	76.23%	42.49%
20	$12.07	76.34%	39.05%

TABLE 7.1: Bermudan option method vs. traditional hedging method: Using the traditional cost of hedging.

Using the results of this chapter, the solution to the max-min problem (7.20) is given by solving a quantile hedging problem for a European put option at the maturity. Let us consider a GMDB contract with the parameters given as above, and maturity time $T = 20$ years. Figure 7.1 plots the maximum probability of the worst-case scenario versus the available initial capital, where the given initial capital is a percentage of the initial cost of superhedging of the embedded Bermudan option.

In Table 7.1 and Table 7.2, we compare the numerical results obtained

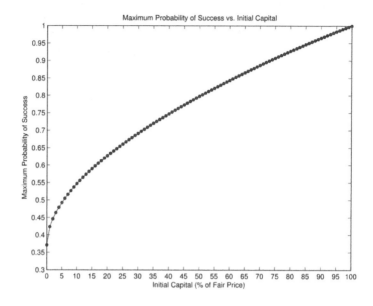

FIGURE 7.1: Maximum probability of success for a GMDB contract using the Bermudan option method.

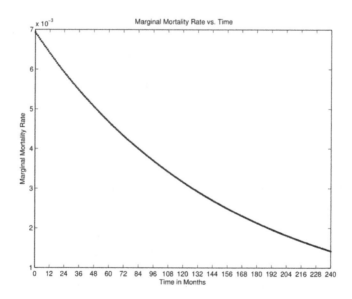

FIGURE 7.2: Marginal mortality rates used for calculating the cost of the hedge portfolio under the traditional method.

Maturity T (years)	Total Initial Capital	Bermudan Option Method	Traditional Hedging Method
5	$4.23	80.32%	53.88%
10	$10.91	86.57%	46.97%
15	$17.55	91.06%	42.49%
20	$23.54	94.28%	39.05%

TABLE 7.2: Bermudan option method vs. traditional hedging method: Using the Bermudan option. fair price

for the quantile hedging problem of a GMDB contract using the Bermudan option method with the results from the traditional hedging method.

We know from Section 7.1 that the traditional hedging method for a GMDB contract suggests hedging a sequence of embedded put options $\{(K-S_t)^+ \mid t \in \{1,...,T\}\}$ starting with the initial capital $_{t-1}p_{x1}q^d_{x,t-1}E^*[(K-S_t)^+]$, for $t = 1, 2, ..., 12T$. Under this hedging method, the corresponding problem to the max-min problem (7.2) is simply the minimum value of $12T$ quantile hedging problems (maximum probability of success) for the $12T$ embedded put option described as above. More precisely, for each $t = 1, 2, ..., 12T$, we solve the following problems separately:

$$P_t := \sup_{\pi \in \mathcal{A}^{\mathbb{F}}(v_0^t)} P(V_t^{v_0^t,\pi} \geq (K-S_t)^+), \qquad (7.49)$$

where $v_0^t := {}_{t-1}p_{x1}q^d_{x,t-1}E^*[(K-S_t)^+]$ is the available cost of hedging in each case. Then, we take the minimum of P_t over $t \in \{1, 2, ..., 12T\}$, i.e.,

$$\min_{t \in \{1,2,...,12T\}} P_t \qquad (7.50)$$

In this framework, this definition of maximizing the worst-case scenario of a successful hedge makes sense, because the $12T$ embedded put options are hedged separately with separate allocated initial capital. In Table 7.1, starting with the price calculated by the traditional method from Section 7.1 equation (7.10), we compare the maximal probability of worst-case scenario hedge for the above two methods. For instance, for $T = 5$ years, the total initial capital is calculated as follows:

$$\sum_{t=1}^{60} {}_{t-1}p_{x1}q^d_{x,t-1}E^*[(K-S_t)^+] = 2.34$$

For the Bermudan option method, we use Theorem 7.3; and for the traditional method, (7.49) and (7.50) are utilized to get the numerical results. As Table 7.1 shows, our method demonstrates better performance compared to

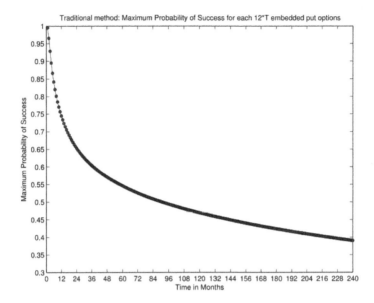

FIGURE 7.3: Maximum probability of success for each individual 12T embedded put option in GMDB contract. In this case, time to maturity T is equal to 20 years.

the traditional hedging method. To explain the underperformance of the second method, we should investigate the way that a GMDB contract is priced by using the marginal mortality rates $_{t-1}p_x {}_1q^d_{x,t-1}$ in equation (7.10). As t approaches the contract maturity $12T$, the survival probability $_{t-1}p_x$ decreases and consequently $_{t-1}p_x {}_1q^d_{x,t-1}$ becomes smaller. This implies that the initial capital allocated to the hedge embedded put option $(K - S_t)^+$ shrinks as t is approaching $12T$. By this method, starting with the initial capital $_{12T-1}p_x {}_1q^d_{x,12T-1}E^*\big[(K - S_{12T})^+\big]$, the lowest probability of a successful hedge always occurs in hedging $(K - S_{12T})^+$. See Figure 7.2 and Figure 7.3 for an illustration of this discussion. This method of hedging is also problematic in the cases where the policyholder dies close to the maturity and at the same time the guarantee is in the money, this can potentially result in drastic losses.

In contrast, we consider a GMDB contract as a Bermudan option that can be (randomly) exercised at any date in $1, 2, ..., 12T$. The superhedging method used to hedge this Bermudan-style option contract determines the initial cost of the hedge such that all possible payoffs are covered. This method of hedging does not underestimate the GMDB payoffs as t is increasing.

Table 7.2 exhibits a similar comparison as in Table 7.1. However, in this case, the initial cost used for hedging is the initial cost of the hedge portfolio for the corresponding Bermudan option multiplied by $P(T(x) \leqslant T)$. Using equation (7.11), this initial capital, second column of Table 7.2, is then distributed into $12T$ marginal capitals as the initial capitals required for problem (7.49)–(7.50).

For example, for the maturity time $T = 5$ years, the initial capital is calculated as follows:

$$P(T(x) \leqslant 5) E^*\left[(K - S_5)^+\right] = 34.48\% \times 12.27 = 4.23$$

Then $4.23 is split among the $60 = 12 \times 5$ put options $(K - S_t)^+$ as the following:

$$4.23 = \sum_{t=1}^{60} {}_{t-1}p_x 1 q^d_{x,t-1} E^*\left[(K - S_5)^+\right] = \sum_{t=1}^{60} {}_{t-1}p_x 1 q^d_{x,t-1} \times 12.27 ,$$

. where ${}_{t-1}p_x 1 q^d_{x,t-1}$ is given by the life table, available for instance from Hardy (2003).

Notice that the initial capitals allocated for the last embedded put option in both Table 7.1 and Table 7.2 are equal to ${}_{12T-1}p_x 1 q^d_{x,12T-1} \times E^*[(K - S_{12T})]$. On the other hand, using the traditional hedging method, the worst-case scenario always occurs in the month before the maturity of the contract. Therefore, the last columns of Table 7.1 and Table 7.2 are the same.

Bibliographic Remarks

Chapter 1

In the chapter, the notions and facts from the theory of stochastic processes needed in what follows are presented. We provide facts from the general theory of stochastic processes and information about Wiener, Poisson, and Levy processes, together with a short sketch of the stochastic calculus of semimartingales. These materials are reflected in many books; see, for example, [Protter (2003)], [Cont and Tankov (2004)], [Cohen and Elliott (2015)]. Our exposition of the most important ingredients, for this work of financial markets and option pricing theory (completeness and incompleteness, hedging and superhedging, methodology finding for martingale measures, etc.) follows [Melnikov et al. (2002)]. We introduce in this chapter equity-linked life insurance contracts as innovative insurance instruments to combine both financial and insurance risks and incorporate a new randomness in actuarial calculations, stemming from the financial randomness of insurance guarantees. These contracts have been studied since the middle of the 1970s, and the first quantitative treatment of such innovative insurance products were given by [Brennan and Schwartz (1976)], [Brennan and Schwartz (1979)] and [Boyle and Schwartz (1977)]. In these papers the traditional actuarial principle of equivalence was transformed so that the Black–Scholes formula (see [Black and Scholes (1973)], [Merton (1973)]) appeared in the actuarial science context. Later, in the papers by [Delbaen (1986)], [Bacinello and Ortu (1993)], [Aase and Pearson (1994)], [Nielsen and Sandmann (1995)], and
[Ekern and Persson (1996)] the pricing problem of equity-linked life insurance contracts was analyzed with the help of martingale measures. In some sense, the monograph by [Hardy (2003)] can be regarded as a result of studies in the area for this period. In the chapter, we present a leading idea to investigate such contracts based on partial/imperfect hedging techniques developed in the last two decades in mathematical finance. The essence of our approach is partial/imperfect hedging: quantile hedging, efficient hedging, CVaR hedging (see [Föllmer and Leukert (1999)], [Föllmer and Leukert (2000)], [Spivak and Cvitanic (1999)],
[Rockafellar and Uryasev (2002)], [Melnikov and Smirnov (2012)]), developed for different financial markets and for deterministic and stochastic insurance

guarantees. Equity-linked life insurance now is a rather wide area of research and applications. Therefore, it is difficult to include all exploited methods without loss of a homogeneous character of the book. This is a reason why the well-developed approaches (quadratic hedging and utility maximization) are not presented here (see, for example, [Moeller (1998)], [Young (2003)], [Biagini and Schreiber (2013)]).

Chapter 2

In this chapter we study the pricing/hedging problem for equity-linked life insurance contracts by means of quantile hedging in the Black–Scholes model. Quantile hedging in mathematical finance was developed at the end of 1990s (see [Föllmer and Leukert (1999)], [Spivak and Cvitanic (1999)]. The main success was achieved with the help of the fundamental Neyman-Pearson lemma (see, [Cvitanic and Karatzas (2001)]). We study equity-linked products with constant and flexible guarantees. Quantile hedging as a natural method of pricing of such contracts was proposed in [Melnikov (2004a)]. It was shown already there that quantile prices and hedges contain, as a natural ingredient, the Black–Scholes [Black and Scholes (1973)], [Merton (1973)] formula (constant guarantee) and the Margrabe [Margrabe (1978)] formula (flexible stochastic guarantee). Our exposition here follows [Melnikov et al. (2005)], [Melnikov and Romaniuk (2006)] for the standard Black–Scholes setting and using standard mortality tables from [Bowers et al. (1997)] or
[McGill et al. (2004)]. More general models with stochastic interest rates and with transaction costs were considered in [Gao et al. (2011)],
[Melnikov and Tong (2011)], and [Melnikov and Tong (2014b)]. We note also the paper by [Melnikov and Romaniuk (2006)], where the Lee–Carter mortality model was exploited to improve quantile hedging analysis, and the paper by [Klusik and Palmowski (2011)] where the problem of quantile hedging was developed for a bigger class of contracts. chapter 2 contains two important lemmas, Lemma 2.2 and Lemma 2.3. The first lemma and its proof can be found in [Föllmer and Leukert (1999)] and in [Melnikov (2011)], Lemma 4.2. The second lemma and the corresponding proof are given in [Melnikov (2011)], Lemma 4.1. We also note that the paper by [Melnikov and Romaniuk (2008)] contains its multivariate version as Theorem 4.1.

Chapter 3

In this chapter, equity-linked life insurance contracts are studied by means of efficient hedging. This type of imperfect hedging was developed by [Föllmer and Leukert (2000)] with the help of loss functions. They connected this problem to the problem of finding an optimal decision rule. Then the corresponding optimal decision was derived from the fundamental Neyman-Pearson lemma.

The efficient hedging approach was proposed for pricing of equity-linked life insurance contracts in [Melnikov (2004b)], and later was developed in [Kirch and Melnikov (2005)], [Melnikov and Romaniuk (2008)], and [Melnikov and Skornyakova (2011)]. Our exposition here follows to the last two papers. Let us note that an efficient hedging problem was also studied for the Black–Scholes model with stochastic interest rate in [Melnikov and Tong (2014a)].

Chapter 4

In Chapter 4, the quantile hedging method is applied to the Black–Scholes model, two-factor diffusion and jump-diffusion models. Besides getting reasonable theoretical results for prices, hedging ratios and survival probabilities, we demonstrate the pricing differences between these three models as well as between equity-linked life insurance contracts with constant and stochastic guarantees. Exposition is given in a way that is most convenient for practitioners. To estimate the model parameters, we use [Andersen and Andreasen (2000)], [Tsay (2002)], [Mancini (2004)]. We provide here a convenient risk management scheme adapted to quantile hedging and formulate the main steps of quantile risk management. Our exposition is based on the talk by [Melnikov and Skornyakova (2005)]: $http://www.actuaries.ca/meetings/stochastic-investment/2006/pdf/1203\_v.2.pdf$.

Chapter 5

In Chapter 5, we study the imperfect hedging problem using one of the most important risk measures CVaR, see, for example,
[Acerbi and Tasche (2002)]. Our approach is based on
[Föllmer and Leukert (1999)], and [Rockafellar and Uryasev (2002)]. We de-

rive explicit formulas for CVaR hedging and apply these findings to pricing of equity-linked life insurance contracts for the Black–Scholes model and a regime-switching telegraph model. Moreover, we demonstrate how CVaR hedging can be exploited for the needs of financial regulation. Our exposition follows two papers [Melnikov and Smirnov (2012)], and [Melnikov and Smirnov (2014)]. All necessary details on the regime-switching telegraph model can be found in [Ratanov and Melnikov (2008)].

Chapter 6

In this chapter, the problem of efficient hedging is studied for (multiple) defaultable claims, see, for example, [Kusuoka (1999)],
[Bielecki and Rutkowski (2004)], [Coculescu et al. (2012)]. Key results from [Föllmer and Leukert (2000)] are used to solve this problem. Usually defaultable markets are incomplete. Therefore, a superhedging approach which is based on optional decomposions of positive supermartingales is exploited here, see [El Karoui and Quenez (1995)], [Kramkov (1996)]. Another key component of our solution to the problem comes from the convex duality approach of [Cvitanic and Karatzas (2001)]. Combining all these methods together with papers by [Nakano (2011)] and [Jiao and Pham (2011)] and using the Frechet and Gateaux derivatives technique we give the full solution to the efficient hedging problem under much better conditions in comparison with [Aguilar(2008)] and [Nakano (2011)].These findings are applied to pricing of equity-linked life insurance contracts. Our exposition here is adapted from the paper by [Melnikov and Nosrati (2015)].

Chapter 7

In this chapter, we study a GMDB contract (see also the OBPI contract introduced by [Leland and Rubinstein (1976)]) in the framework of the Black–Scholes model. The main task is to solve the quantile hedging problem for such a contract under an initial capital constraint. To do it, we progressively enlarge the underlying filtration with the filtration generated by the survival process. The key viewpoint to the GMDB contract as a Bermudan option on the probability space equipped with the enlarged filtration. It gives the possibility to apply here a superhedging approach inspired by [Schweizer (2002)] for pricing of Bermudan options. To realize it in detail we use [Föllmer and Leukert (1999)],

[Cvitanic (2000)], [Pham (2010)] and [Nakano (2011)]. Our exposition follows the paper by [Melnikov and Nosrati (2017)].

Bibliography

[Aase and Pearson (1994)] Aase, K. and S. Persson, 1994. *Pricing of unit-linked insurance policies.* Scandinavian Actuarial Journal 1: 26–52.

[Aguilar(2008)] Aguilar, E. T., 2008. *American options in incomplete markets: Upper and lower Snell envelopes and robust partial hedging.* PhD thesis, Humboldt University of Berlin, Germany.

[Acerbi and Tasche (2002)] Acerbi, C. and D. Tasche, 2002. *On the coherence of expected shortfall.* Journal of Banking and Finance 26(7): 1487–1503.

[Andersen and Andreasen (2000)] Andersen, L. and J. Andreasen, 2000. *Jump-diffusion processes: Volatility smile fitting and numerical methods for option pricing.* Review of Derivatives Research 4: 231–262.

[Bacinello and Ortu (1993)] Bacinello, A. R. and F. Ortu, 1993. *Pricing of unit-linked life insurance with endogenous minimum guarantees.* Insurance: Mathematics and Economics 12: 245–257.

[Biagini and Schreiber (2013)] Biagini, F. and I. Schreiber, 2013. *Risk-minimization for life insurance liabilities.* SIAM Journal on Financial Mathematics 4: 243–264.

[Bielecki and Rutkowski (2004)] Bielecki, T. R. and M. Rutkowski, 2004. *Credit risk: Modeling, valuation and hedging.* Springer Finance, Berlin.

[Black and Scholes (1973)] Black, F. and M. Scholes, 1973. *Pricing of options and corporate liabilities.* Journal of Political Economy 81: 637–654.

[Bowers et al. (1997)] Bowers, N. L., H. U. Gerber, G. C. Hickman, D. A. Jones, and C. J. Nesbit, 1997. *Actuarial Mathematics.* The Society of Actuaries. Schaumburg, Illinois.

[Boyle and Hardy (1997)] Boyle, P. P. and M. Hardy, 1997. *Reserving for maturity guarantees: Two approaches.* Insurance: Mathematics and Economics 21: 113–127.

[Boyle and Schwartz (1977)] Boyle, P. P. and E. S. Schwartz, 1977. *Equilibrium prices of guarantees under equity-linked contracts.* Journal of Risk and Insurance 44: 639–680.

[Brennan and Schwartz (1976)] Brennan, M., and E. S. Schwartz, 1976. *The pricing of equity-linked life insurance policies with an asset value guarantee.* Journal of Financial Economics 3: 195–213.

[Brennan and Schwartz (1979)] Brennan, M., and E. S. Schwartz, 1979. *Alternative investment strategies for the issuers of equity-linked life insurance with an asset value guarantee.* Journal of Business 52: 63–93.

[Coculescu et al. (2012)] Coculescu, D., M. Jeanblanc, and A. Nikeghbali, 2012. *Default times, non-arbitrage conditions and change of probability measures.* Finance and Stochastics 16: 513–535.

[Cohen and Elliott (2015)] Cohen, S. N. and R. J. Elliott, 2015. *Stochastic calculus and applications.* Springer, New York, Heidelberg, Dordrecht, London.

[Cont and Tankov (2004)] Cont, R. and P. Tankov, 2004. *Financial modelling with jump processes.* Chapman and Hall/CRC, Boca Raton, Florida.

[Cvitanic (2000)] Cvitanic, J., 2000. *Minimizing expected loss of hedging in incomplete and constrained markets.* SIAM Journal of Control Optimization 38: 1050–1066.

[Cvitanic and Karatzas (2001)] Cvitanic, J. and I. Karatzas, 2001. *Generalized Neyman–Pearson lemma via convex duality.* Bernoulli 7: 79–97.

[Delbaen (1986)] Delbaen, F., 1986. *Equity-linked policies.* Bulletin Association Royal Actuaries Belges 80: 33–52.

[Ekern and Persson (1996)] Ekern, S. and S. Persson, 1996. *Exotic unit-linked life insurance contracts.* The Geneva Papers on Risk and Insurance Theory 21: 35–63.

[El Karoui and Quenez (1995)] El Karoui, N. and M. Quenez, 1995. *Dynamic programming and pricing of contingent claims in an incomplete market.* SIAM Journal of Control and Optimization 33: 29–66.

[Föllmer and Leukert (1999)] Föllmer, H., and P. Leukert, 1999. *Quantile Hedging.* Finance and Stochastics 3: 251–273.

[Föllmer and Leukert (2000)] Föllmer, H., and P. Leukert, 2000. *Efficient hedging: Cost versus short-fall risk.* Finance and Stochastics 4: 117–146.

[Gao et al. (2011)] Gao, Quansheng, Ting He and Chi Zhang, 2011. *Quantile hedging for equity-linked life insurance contracts in a stochastic interest rate economy*. Economic Modelling 28: 147–156.

[Hardy (2003)] Hardy, M., 2003. *Investment guarantees: Modeling and risk management for equity-linked life insurance*. John Wiley & Sons, Inc., Hoboken, New Jersey.

[Jiao and Pham (2011)] Jiao, Y. and H. Pham, 2011. *Optimal investment with counterparty risk: A default-density model approach*. Finance and Stochastics 15(4): 725–753.

[Kirch and Melnikov (2005)] Kirch, M., and A. Melnikov, 2005. *Efficient hedging and pricing of life insurance policies in a jump-diffusion model*. Stochastic Analysis and Applications 23: 1213–1233.

[Klusik and Palmowski (2011)] Klusik, P. and Z. Palmowski, 2011. *Quantile hedging for equity-linked contracts*. Insurance: Mathematics & Economics 48: 280–286.

[Kramkov (1996)] Kramkov, D. O., 1996. *Optional decomposition of supermartingales and hedging contingent claims in incomplete security markets*. Probability Theory and Related Fields 105: 459–479.

[Kusuoka (1999)] Kusuoka, S., 1999. *A remark on default risk models*. Advances in Mathematical Economics 1: 69–82.

[Leland and Rubinstein (1976)] Leland, H. E. and M. Rubinstein, 1976. *The evolution of portfolio insurance. In Luskin, D. L. (Ed.), Portfolio insurance: A guide to dynamic hedging*. John Wiley & Sons, Inc., New York.

[Mancini (2004)] Mancini, C., 2004. *Estimation of the characteristics of the jumps of a general Poisson-diffusion model*. Scandinavian Actuarial Journal 1: 42–52.

[Margrabe (1978)] Margrabe, W., 1978. *The value of an option to exchange one asset to another*. Journal of Finance 33: 177–186.

[McGill et al. (2004)] McGill, D., K. Brown, J. Haley, and S. Schieber, 2004. *Fundamentals of Private Pensions*. Oxford University Press. 8th Edition.

[Melnikov (2004a)] Melnikov, A. V., 2004. *Quantile hedging of equity-linked life insurance policies*. Doklady Mathematics 69(3): 428–430.

[Melnikov (2004b)] Melnikov, A. V., 2004. *Efficient hedging of equity-linked life insurance policies*. Doklady Mathematics 69(3): 462–464.

[Melnikov (2011)] Melnikov, A., 2011. *Risk Analysis in Finance and Insurance*. 2nd Edition, Chapman and Hall / CRC, Boca Raton, Florida.

[Melnikov et al. (2002)] Melnikov, A. V., S. N. Volkov, and M. L. Nechaev, 2002. *Mathematics of financial obligations*. American Mathematical Society, Providence, RI.

[Melnikov et al. (2005)] Melnikov, A. V., Y. V. Romaniuk and V. S. Skornyakova, 2005. *The Margrabe formula and quantile hedging of life insurance policies*. Doklady Mathematics 71(1): 31–34.

[Melnikov and Nosrati (2015)] Melnikov, A. and A. Nosrati, 2015. *Efficient hedging for defaultable securities and its application to equity-linked life insurance contracts*. International Journal of Theoretical and Applied Finance 18(7): DOI: 10.1142/S0219024915500478.

[Melnikov and Nosrati (2017)] Melnikov, A. and A. Nosrati, 2017. *Bermudan options and connections to equity-linked life insurance contracts*. Insurance: Mathematics and Economics (submitted).

[Melnikov and Romaniuk (2006)] Melnikov, A. and Y. Romaniuk, 2006. *Evaluating the performance of Gompertz, Makeham and Lee-Carter mortality models for risk management with unit-linked contracts*. Insurance: Mathematics and Economics 39(3): 310–329.

[Melnikov and Romaniuk (2008)] Melnikov, A. and Y. Romaniuk, 2008. *Efficient hedging and pricing of equity-linked life insurance contracts on several assets*. International Journal of Theoretical and Applied Finance 11(3): 295–323.

[Melnikov and Skornyakova (2005)] Melnikov, A. and V. Skornyakova, 2005. *Quantile hedging and its application to life insurance*. Statistics and Decisions 23(4): 301–316.

[Melnikov and Skornyakova (2011)] Melnikov, A. and V. Skornyakova, 2011. *Efficient hedging as risk management methodology in equity-linked life insurance. In Nota, G. (Ed.), Risk management trends*. InTech, Rijeka, Croatia: 149–166.

[Melnikov and Smirnov (2012)] Melnikov, A. and I. Smirnov, 2012. *Dynamic hedging of conditional value-at-risk*. Insurance: Mathematics and Economics 51(1): 182–190.

[Melnikov and Smirnov (2014)] Melnikov, A. and I. Smirnov, 2014. *Option pricing and CVaR hedging in the regime-switching telegraph market model. In Silvestrov, D. and A. Martin-Loef (Eds), Modern problems in insurance mathematics*. Springer, Cham, Heidelberg, New York, Dordrecht, London: 365–378.

[Melnikov and Tong (2011)] Melnikov, A. and S. Tong, 2011. *Quantile hedging for equity-linked life insurance contracts with stochastic interest rate.* Procedia Systems Engineering (Elsevier) 4: 9–24.

[Melnikov and Tong (2014a)] Melnikov, A. and S. Tong, 2014. *Valuation of finance/insurance contracts: Efficient hedging and stochastic interest rate modeling.* Risk and Decision Analysis 5: 23–41.

[Melnikov and Tong (2014b)] Melnikov, A. and S. Tong, 2014. *Quantile hedging on equity-linked life insurance contracts with transaction costs.* Insurance: Mathematics and Economics 55: 250–260.

[Merton (1973)] Merton, R. C., 1973. *Theory of rational option pricing.* Bell Journal of Economics and Management Science 4: 141–183.

[Merton (1976)] Merton, R. C., 1976. *Option pricing when underlying stock returns are discontinuous.* Journal of Financial Economics 3: 125–144.

[Moeller (1998)] Moeller, T., 1998. *Risk-minimizing hedging strategies for unit-linked life-insurance contracts.* Astin Bulletin 28: 17–47.

[Nakano (2011)] Nakano, Y., 2011. *Partial hedging for defaultable claims.* Advances in Mathematical Economics 14: 127–145.

[Nielsen and Sandmann (1995)] Nielsen, J. A. and K. Sandmann, 1995. *Equity-linked life insurance: A model with stochastic interest rate.* Insurance: Mathematics and Economics 16: 225–253.

[Pham (2010)] Pham, H., 2010. *Stochastic control under progressive enlargement of filtrations and applications to multiple defaults risk management.* Stochastic Processes and their Applications 120: 1795–1820.

[Protter (2003)] Protter, P. E., 2003. *Stochastic integration and differential equations.* Springer, Berlin.

[Ratanov and Melnikov (2008)] Ratanov, N. and A. Melnikov, 2008. *On financial markets based on telegraph processes.* Stochastics: An International Journal of Probability and Stochastic Processes 80(2): 247–268.

[Rockafellar and Uryasev (2002)] Rockafellar, R.T. and S. Uryasev, 2002. *Conditional value-at-risk for general loss distributions.* Journal of Banking and Finance 26: 1443–1471.

[Schweizer (2002)] Schweizer, M., 2002. *On Bermudan options.* Advances in Finance and Stochastics. Essays in Honor of Dieter Sondermann: 257–269.

[Spivak and Cvitanic (1999)] Spivak, G. and J. Cvitanic, 1999. *Maximizing the probability of a perfect hedge*. The Annals of Applied Probability 9: 1303–1328.

[Tsay (2002)] Tsay, R. S., 2002. *Analysis of Financial Time Series*. John Wiley & Sons, Inc.

[Young (2003)] Young, V. R., 2003. *Equity-indexed life insurance: Pricing and reserving using the principle of equivalent utility*. North American Actuarial Journal 7: 68–86.

Index

Admissible strategy, 25, 34, 53, 83, 109
American option, 163, 167, 168
Arbitrage, 13, 32, 65, 128, 166
Asset, 13, 21, 31, 32, 53, 125, 128

Backward induction, 169
Balance equation, 18, 37, 67, 78
Bank account, 32, 43, 63, 137
Bermudan option, 163
Black–Scholes differential equation, 18
Black–Scholes formula, 18, 20, 21, 97, 117, 178
Black–Scholes model, 17, 23, 53, 90, 91, 115, 119, 121, 123, 137
Brennan–Schwartz price, 22, 27, 35, 66, 77, 93, 122
Brownian motion, 2

Canonical decomposition, 10, 12
Characteristic equation, 39, 57, 71, 79, 87, 96
Compensator, 8
Compensator of the measure, 10
Complete market, 14, 34, 109, 137
Conditional Value-at-Risk, 107–109, 112, 119, 129, 131, 134
Constant guarantee, 159, 163
Consumption, 14, 15
Consumption process, 170, 173
Contingent claim, 14, 15, 17, 23, 24, 34, 54, 65, 83, 92
Continuous process, 2, 4, 6, 173
Correlation, 64
Cumulative gains, 13
Cumulative losses, 13

CVaR hedging, 107, 115, 124, 125
Default risk, 28
Default time, 137–139, 158, 161
Defaultable claim, 137
Density, 3, 19, 27, 33
Derivative, 20, 111, 144, 153
Derivative security, 20
Deterministic guarantee, 23, 26
Diffusion model, 77
Diffusion process, 3, 5
Doob–Meyer decomposition, 8

Efficient hedging, 20, 53, 64, 143
Embedded option, 78, 93
Enlarged filtration, 152, 155, 163, 167, 181
Equity-linked insurance, 47, 100
Equity-linked life insurance, 90
Equity-linked life insurance contracts, 21, 26, 27, 53, 64, 115, 137, 158, 163
Esscher transform, 12
European call option, 134
European option, 14, 84, 164, 169, 175, 177
European put option, 181

Fair price, 15, 28, 47, 49, 66, 129, 183
Financial market, 13, 15, 23, 26, 79, 84, 91, 96–98
Financial risk, 21, 37, 55, 67, 95, 96
Flexible guarantee, 32, 46, 90, 93, 99
Force of mortality, 161, 167

Geometrical Brownian motion, 64
Girsanov exponential, 4, 11
Girsanov theorem, 4, 17, 65, 115

Index

GMDB, 163–168, 170, 175, 179, 181–183
GMMB, 164
Guarantee, 21, 23, 53, 64, 77, 165

Hazard rate function, 161, 167
Hedge, 14, 99, 109
Hedging, 14, 15, 23, 84, 90, 114
Hedging ratio, 27

Incomplete market, 15, 20, 23, 137, 141, 159, 167
Insurance risk, 21, 55, 92
Interest rate, 24, 30, 32, 33, 42, 62, 77, 82, 115, 126, 137, 165
Investment strategy, 13
Ito formula, 4, 5
Ito process, 4

Jump-diffusion model, 18, 82, 85, 87, 90, 91, 100

Kunita–Watanabe decomposition, 9

Law of large numbers, 95, 159, 166
Levy process, 11, 12
Life insurance, 21, 22, 36, 82, 84, 163
Life table, 47, 48, 50, 80, 123, 158, 185
Local density, 3, 11, 16, 83
Local martingale, 8–11
Loss function, 20, 53, 56, 62, 109, 140
Loss ratio, 61, 84

Margrabe formula, 32, 67, 69, 84
Market, 13
Martingale, 5, 6, 8
Martingale measure, 12, 24
Martingale representation, 5
Maturity guarantee, 21
Maximal successful hedging set, 25, 90, 95, 99
Mean-variance hedging, 20
Measurable process, 4
Minimum guarantee, 166, 181
Modified claim, 35, 43, 72, 144, 159, 176

Mortality table, 22
Multiple contract, 46
Multiple defaults, 137

Neyman–Pearson lemma, 23, 24
Normal distribution, 28

Option pricing, 15, 127
Optional decomposition, 12, 141, 151
Optional process, 7

Partial hedging, 23, 107
Perfect hedge, 14, 25
Perfect hedging, 20
Poisson distribution, 5, 85
Poisson martingale, 5
Poisson process, 5
Portfolio, 13, 32, 82
Predictable characteristics, 10
Predictable process, 6, 26, 32, 82
Predictable stopping time, 7
Probability measure, 2, 3, 23
Progressive measurability, 1
Pure endowment contract, 35, 65, 77

Quadratic characteristic, 8, 9
Quantile hedge, 25, 81, 83
Quantile hedging, 20, 23, 77
Quasi-continuous process, 7

Radon–Nikodym density, 3
Random measure, 9
Regime-switching telegraph market model, 125
Replicating strategy, 14, 178
Right-continuous process, 6
Risk aversion, 72
Risk-measure, 94, 107
Risk-neutral measure, 27, 33
Risky asset, 21

Segregated fund, 21
Self-financing strategy, 14, 24, 92, 108, 140
Semimartingale, 6, 9, 10
Shortfall, 53

Shortfall risk, 53
Stochastic analysis, 1, 6
Stochastic differential equation, 5
Stochastic guarantee, 32, 46, 90, 93
Stochastic integral, 3, 5, 9
Stochastic interval, 7
Stochastic process, 1, 4
Stock, 21, 64, 77
Stopping time, 7, 163
Submartingale, 8
Successful hedging, 20, 27, 28, 34, 81, 83
Successful hedging set, 24, 83, 94
Superhedge, 137, 159
Superhedging, 137, 141, 158, 164
Supermartingale, 8
Survival probability, 21, 26, 28, 79, 93, 121, 123, 158
Survival process, 138

Telegraph process, 127
Trading strategy, 26, 32, 82, 140

Wiener measure, 2
Wiener process, 2